METHODS IN MOLECULAR BIOLOGY

Series Editor
John M. Walker
School of Life and Medical Sciences
University of Hertfordshire
Hatfield, Hertfordshire, AL10 9AB, UK

For further volumes:
http://www.springer.com/series/7651

Clathrin-Mediated Endocytosis

Methods and Protocols

Edited by

Laura E. Swan

*Department of Cellular & Molecular Physiology, University of Liverpool,
Institute of Translational Medicine, Liverpool, UK*

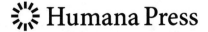 Humana Press

Editor
Laura E. Swan
Department of Cellular & Molecular Physiology
University of Liverpool
Institute of Translational Medicine
Liverpool, UK

ISSN 1064-3745   ISSN 1940-6029   (electronic)
Methods in Molecular Biology
ISBN 978-1-4939-9374-1   ISBN 978-1-4939-8719-1   (eBook)
https://doi.org/10.1007/978-1-4939-8719-1

This Humana Press imprint is published by the registered company Springer Science+Business Media, LLC part of Springer Nature.
The registered company address is: 233 Spring Street, New York, NY 10013, U.S.A.

Preface

The regulated retrieval of plasma membrane and its cargo proteins is fundamental to a cell's interaction with its environment. In a membrane bristling with transmembrane receptors, at any given moment some of them signaling, some of them silent—the ability to selectively isolate, concentrate, and retrieve specific cargoes can dynamically and locally regulate cell signaling, cell adhesion, and membrane properties.

One of the best-understood modes of regulated membrane retrieval is clathrin-mediated endocytosis (CME): the assemblage of a macromolecular scaffold of membrane-associated adaptor proteins which are progressively "caged" by clathrin triskelia, bud off the plasma membrane, and are cleaved away from the plasma membrane, to release a packet of membrane for endocytic processing and/or recycling.

CME can be extremely efficient: for example, in certain conditions up to 40% of plasma membrane transferrin receptor can be retrieved by CME in a minute. CME can also be very high fidelity: synaptic vesicles retrieved by CME have highly stereotypical diameters, as compared to CME of other cargoes, which allows refilling of synaptic vesicles with uniform amounts of neurotransmitter, and thus scalable neurotransmission at synapses distant from the cell body. Meanwhile CME retains a high degree of flexibility, including large or multiple cargoes and a wide range of vesicle sizes when required.

What we now know as CME was first noted nearly 60 years ago in a series of electron micrographs by Roth and Porter, where they noted a regular bristle-like or lattice-like array on juxta-membrane vesicles in mosquito oocytes, which they astutely noted gave way to "naked" packages of similarly sized membrane further within the cell, suggesting a transient coat-like structure which they proposed might assist the internalization of membrane. Once purified to near homogeneity by Barbara Pierce in 1975, we learnt that this coat was made predominantly of a single protein: clathrin heavy chain. Subsequent experiments complemented by beautiful electron microscopy studies by Heuser, Kanaseki and Kadota, Ungewickel and Branton among others showed that clathrin heavy chain associated as a triskelion reinforced and "buttressed" by shorter clathrin light chains, which naturally assemble into a combination of hexagon and pentagon geometries allowing to form a curved "basket" around membrane. Keen utilized biochemical fractionations to discover its function and assemblage was potentiated by what he called assembly polypeptides (APs), adaptor proteins that link clathrin to the plasma membrane. Since then a number of extremely fine studies (some of them contributed by authors in this book) have started to pick this structure apart to understand how CME can dynamically respond to plasma membrane status and signaling.

Studies in plasma membrane clathrin have revealed several other, useful general principles, first among them that the clathrin coat itself is used on other cellular membranes, linked by specific AP complexes (AP-1 at the trans-golgi and recycling endosome, AP-2 at plasma membrane, AP-3 at tubular endosomes and involved in the biogenesis of lysosome-related organelles and release of exocytic organelles) and similar complexes (AP-4 and AP-5) which, having the same structure as AP1-3, might be expected to be associated with a clathrin-like coat. This has yet to be identified for AP-4, but intriguing possibilities suggest that AP-5 complex may associate with a protein SPG11 (mutated in hereditary spastic paraplegias), which shares several structural features with clathrin. Indeed, it has subsequently been found that clathrin-like coat proteins are responsible for targeted transport of membrane packets across the cell, allowing transmembrane and secreted proteins to be condensed and fine-regulated as they pass through membranous compartments.

While the process of CME is fundamental, it is also extremely responsive to context (as it must be, to accomplish the myriad of tasks it must perform simultaneously in a single cell); consequently, since these seminal studies there has been no shortage of experiments to do: isolating and identifying coat components, accessory factors, probable geometries, and recruitment and dispersal cues has led to a many-pronged assault on this topic, which is bearing fruit to this day. The advent of high resolution microscopy, semi-automated microscopy, and image processing bioinformatics approaches has seen an explosion in high-throughput studies where many thousands of clathrin assembly or budding steps can be monitored or modified (some of which we present here), which has allowed for a subtle understanding of context-dependent mechanisms of CME and factors, such as membrane tension, lipid content, and actin nucleation, which alter CME kinetics and fidelity in various systems. As such, there is still a lot to discover about how and why clathrin coats form, what causes them to commit to a certain geometry, and how the cell "decides" what speed this process takes place—all of which have implications for how the cell signals and interacts with its environment.

On one hand, because of the clear importance of CME to neurotransmission, cell signaling through growth factors, and cell membrane properties, CME is worth in and of itself of the level of attention and study that has been devoted to it for over half a century. Second, due to being essentially confined in two dimensions (although as recent studies with light-sheet microscopy show, CME has its own regulatory mechanisms on the dorsal surfaces of cells), and being stimulus-dependent, CME, of all the coat maturation processes, is the easiest to access and biochemically or visually isolate—allowing systematic, time-resolved studies approaches to imaging, biochemistry, and electron microscopy, which have been valuable instructors for similar processes across the cell.

This book aims to present a methodological introduction to the world of clathrin-mediated endocytosis, suggesting a number of model systems, purifications, reconstitutions, analysis techniques, and systemic alterations of membrane function, which have been pivotal in understanding the role of clathrin and is panoply of interactors, contributed by some of the leading practitioners in the field today. I hope that it will allow a point of entry for both the novice and experienced scientist with new questions to ask. We now in 2018 have an array of arms at our disposal from both big data approaches and fine-grained reconstructions, which are truly beginning to understand how the cell goes about its most important job: knowing when to "listen" to the outside world, and when to specifically turn that signal down. I am extremely obliged to the authors who have contributed to this book for their time and patience.

Liverpool, UK   *Laura E. Swan*

Contents

The original version of this book was revised. The correction is available at
https://doi.org/10.1007/978-1-4939-8719-1_19

Contributors

FRANÇOIS AGUET • *The Broad Institute of MIT and Harvard, Cambridge, MA, USA*

DEVIKA ANDHARE • *Indian Institute of Science Education and Research, Pune, Maharashtra, India*

ONA E. BLOOM • *Department of Physical Medicine and Rehabilitation, The Hofstra North Shore-LIJ School of Medicine, The Feinstein Institute for Medical Research, Manhasset, NY, USA*

TILL BÖCKING • *ARC Centre of Excellence in Advanced Molecular Imaging and EMBL Australia Node in Single Molecule Science, School of Medical Sciences, University of New South Wales, Sydney, Australia*

FRANCESCA BOTTANELLI • *Department of Cell Biology, Yale School of Medicine, New Haven, CT, USA*

EMMANUEL BOUCROT • *Institute of Structural and Molecular Biology, University College London and Birkbeck, London, UK*

STEEVE BOULANT • *Department of Infectious Diseases, Virology, Heidelberg University, Heidelberg, Germany; Schaller Research Group at CellNetworks, Heidelberg, Germany; German Cancer Research Center (DKFZ), Heidelberg, Germany*

BENJAMIN R. CAPRARO • *Department of Cell Biology, Harvard Medical School, Boston Children's Hospital, Boston, MA, USA; Program in Cellular and Molecular Medicine, Boston Children's Hospital, Boston, MA, USA*

YAN CHEN • *Department of Biological Sciences, National University of Singapore, Singapore, Singapore; Centre for Bioimaging Sciences, National University of Singapore, Singapore, Singapore*

EMANUELE COCUCCI • *Division of Pharmaceutics and Pharmaceutical Chemistry, College of Pharmacy and the Comprehensive Cancer Center, The Ohio State University, Columbus, OH, USA*

MICHAEL A. COUSIN • *Centre for Discovery Brain Sciences, Edinburgh Medical School: Biomedical Sciences, University of Edinburgh, Edinburgh, UK*

RUBÉN FERNÁNDEZ-BUSNADIEGO • *Department of Molecular Structural Biology, Max Planck Institute of Biochemistry, Martinsried, Germany*

SILVIA GIOVEDÌ • *Department of Experimental Medicine, University of Genova, Genoa, Italy*

SARAH L. GORDON • *Florey Institute of Neuroscience and Mental Health, The University of Melbourne, Parkville, VIC, Australia*

VOLKER HAUCKE • *Leibniz-Forschungsinstitut für Molekulare Pharmakologie, Berlin, Germany; Faculty of Biology, Chemistry and Pharmacy, Freie Universität Berlin, Berlin, Germany*

SACHIN S. HOLKAR • *Indian Institute of Science Education and Research, Pune, Maharashtra, India*

TOM KIRCHHAUSEN • *Department of Cell Biology, Harvard Medical School, Boston Children's Hospital, Boston, MA, USA; Program in Cellular and Molecular Medicine,*

Boston Children's Hospital, Boston, MA, USA; Department of Pediatrics, Harvard Medical School, Boston, MA, USA

MICHAEL KRAUß • *Leibniz-Forschungsinstitut für Molekulare Pharmakologie, Berlin, Germany*

ELSA LAUWERS • *Laboratory of Neuronal Communication, Department of Neurosciences, VIB-KU Leuven Center for Brain and Disease Research, Leuven Brain Institute, KU Leuven, Leuven, Belgium*

ERIC B LEWELLYN • *Department of Biology, Lawrence University, Appleton, WI, USA; Department of Biology, St. Norbert College, De Pere, WI, USA*

XUELIN LOU • *Department of Cell Biology, Neurobiology and Anatomy, Medical College of Wisconsin, Milwaukee, WI, USA*

MIRKO MESSA • *Department of Neuroscience and Cell Biology, HHMI, Program in Cellular Neuroscience, Neurodegeneration and Repair, Yale University School of Medicine, New Haven, CT, USA*

YANSONG MIAO • *School of Biological Sciences, Nanyang Technological University, Singapore, Singapore; School of Chemical and Biomedical Engineering, Nanyang Technological University, Singapore, Singapore*

IRA MILOSEVIC • *European Neuroscience Institute (ENI) and University Medical Center Göttingen (UMG), Göttingen, Germany*

JENNIFER R. MORGAN • *The Eugene Bell Center for Regenerative Biology and Tissue Engineering, Marine Biological Laboratory, Woods Hole, MA, USA*

THOMAS J. PUCADYIL • *Indian Institute of Science Education and Research, Pune, Maharashtra, India*

IRIS RAPOPORT • *Department of Cell Biology, Harvard Medical School, Boston Children's Hospital, Boston, MA, USA; Program in Cellular and Molecular Medicine, Boston Children's Hospital, Boston, MA, USA*

STEPHEN J. ROYLE • *Centre for Mechanochemical Cell Biology, Warwick Medical School, Coventry, UK*

ANTÓNIO J. M. SANTOS • *Cell and Developmental Biology Program, Centre for Genomic Regulation (CRG), Barcelona, Spain*

LENA SCHROEDER • *Department of Cell Biology, Yale School of Medicine, New Haven, CT, USA*

KAREN J. SMILLIE • *Centre for Discovery Brain Sciences, Edinburgh Medical School: Biomedical Sciences, University of Edinburgh, Edinburgh, UK*

SRIGOKUL UPADHYAYULA • *Department of Cell Biology, Harvard Medical School, Boston Children's Hospital, Boston, MA, USA; Program in Cellular and Molecular Medicine, Boston Children's Hospital, Boston, MA, USA; Department of Pediatrics, Harvard Medical School, Boston, MA, USA*

PATRIK VERSTREKEN • *Laboratory of Neuronal Communication, Department of Neurosciences, VIB-KU Leuven Center for Brain and Disease Research, Leuven Brain Institute, KU Leuven, Leuven, Belgium*

RYLIE B. WALSH • *The Eugene Bell Center for Regenerative Biology and Tissue Engineering, Marine Biological Laboratory, Woods Hole, MA, USA*

LAURA A. WOOD • *Centre for Mechanochemical Cell Biology, Warwick Medical School, Coventry, UK; Department of Biology and Biochemistry, University of Bath, Bath, UK*

MIN WU • *Department of Biological Sciences, National University of Singapore, Singapore, Singapore; Centre for Bioimaging Sciences, National University of Singapore, Singapore, Singapore; Mechanobiology Institute, National University of Singapore, Singapore, Singapore*

JEFFERY YONG • *Department of Biological Sciences, National University of Singapore, Singapore, Singapore; Centre for Bioimaging Sciences, National University of Singapore, Singapore, Singapore*

Chapter 1

Purification of Clathrin-Coated Vesicles from Adult Rat Brain

Silvia Giovedì

Abstract

Here, we describe a purification protocol for isolating clathrin-coated vesicles (CCVs) from adult rat brain by using differential centrifugation coupled with Ficoll–sucrose and D_2O–sucrose density gradient centrifugation and an additional linear sucrose step gradient at the end to separate CCVs from contaminating membranes present in the crude microsomal fraction.

Key words Clathrin-coated vesicles, Subcellular fractionation, Differential centrifugation, Density gradient centrifugation, Rat brain

1 Introduction

Clathrin-mediated endocytosis is a major pathway by which proteins and membrane components are internalized into eukaryotic cells and recycled back to the surface (as in early endosomes and recycling endosomes), or sorted to degradation (as in late endosomes and lysosomes). Clathrin-coated vesicles (CCVs) are intermediates in this variety of selective membrane transport processes that derive from clathrin-coated membrane regions predominantly on the plasma membrane, but also on the *trans*-Golgi network, endosomes, and lysosomes by a process of invagination and fission [1–4]. CCVs have been successfully isolated from many different tissues and cell types. The small and relatively uniform size of CCVs (50–100 nm in diameter; [5]) and the high density of these organelles due to their protein coat (1.20–1.25 mg/mL; [6]) allowed the development of reliable and reproducible protocols for their isolation.

According to modifications of original published procedures [7, 8], Maycox et al. [9] developed a subcellular fractionation procedure that led to the isolation of a near homogeneous population

The original version of this chapter was revised. A correction to this chapter can be found at https://doi.org/10.1007/978-1-4939-8719-1_19

Laura E. Swan (ed.), *Clathrin-Mediated Endocytosis: Methods and Protocols*, Methods in Molecular Biology, vol. 1847, https://doi.org/10.1007/978-1-4939-8719-1_1, © Springer Science+Business Media, LLC, part of Springer Nature 2018

of CCVs from adult rat brain by a series of differential, Ficoll–sucrose and D_2O–sucrose density gradient centrifugation steps. A similar procedure coupled to an additional step involving velocity sedimentation in linear sucrose gradients has been used by Wasiak et al. [10] to purify rat brain CCVs for proteomic characterization.

In this protocol, based on Girard et al. [11], adult rat brains are homogenized in MES buffer at mild acidic pH 6.5 (buffer A) using a glass–Teflon homogenizer. The first step of purification involves two differential centrifugations to generate crude microsomes, which are enriched for CCVs. Next, the CCVs are separated from larger microsomal contaminants by velocity sedimentation through a Ficoll–sucrose solution (12.5%), in which CCVs remain in the supernatant. The CCVs are subsequently separated from less dense membranes by centrifugation through a sucrose cushion (8%) prepared in D_2O. Only dense CCVs are pelleted under these conditions, overcoming contamination with synaptic vesicles [9]. In the final, optional step, the D_2O pellets are fractionated by centrifugation on linear sucrose gradient (20%–50%) to determine the fractions further enriched of CCVs and essentially free of any type of contamination.

Figure 1 schematically outlines the subsequent steps used in the purification procedure of CCVs. It is recommended during the procedure to retain aliquots from different fractions to evaluate CCV enrichment and purity. Equal protein aliquots of the various subcellular fractions should be analyzed by one-dimensional SDS-PAGE and Coomassie Blue staining and/or immunoblotting using specific CCV proteins as a marker. Enrichment for major components of clathrin coat, including clathrin heavy chain (CHC), clathrin light chain (CLC), the α-, β-, γ-, and μ-adaptin subunits of the AP-2 and AP-1 clathrin adaptor complexes, should be easily detectable going from the crude homogenate to the CCV fraction. The purity of CCVs could be also best evaluated by electron microscopy (EM; [12]).

2   Materials

Use deionized distilled water in all recipes and protocol steps.

All prepared solutions, glassware, centrifuge tubes, rotors, and equipment should be precooled, and samples should be kept on ice throughout the preparation, unless otherwise noted.

1. 10 Sprague-Dawley rats (150–200 g each).

2. 10× Buffer A: 1 M 2-(*N*-morpholino)ethanesulfonic acid (MES); 10 mM ethylene glycol bis(β-aminoethylether)-*N,N,N',N'*-tetraacetic acid (EGTA); 5 mM $MgCl_2$. Adjust to pH 6.5 with 10 N NaOH. Store up to 4 weeks at 4 °C.

3. 1× Buffer A: To 400 mL H_2O, add 50 mL of 10× Buffer A. Adjust to pH 6.5 with 1 N NaOH if necessary. Add H_2O to

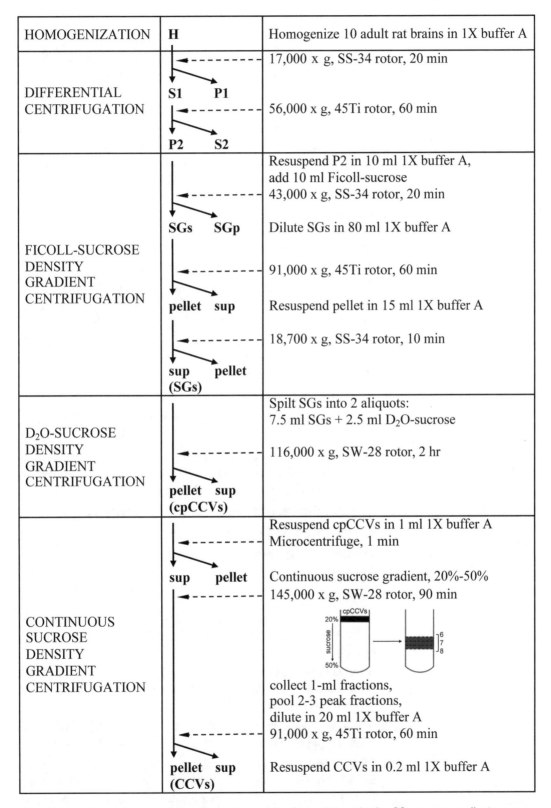

Fig. 1 Scheme for purification of clathrin-coated vesicles from adult rat brain. *SGs* sucrose gradient supernatant, *SGp* sucrose gradient pellet, *cpCCVs* cushion-pellet clathrin-coated vescicles, *CCVs* clathrin-coated vescicles

500 mL. Store up to 2 days at 4 °C. To prevent the appearance of dust or other particles in the sample to be analyzed, pass 1× buffer A through a 0.22-μm filter before use. Add protease inhibitors (aprotinin, benzamidine, leupeptin, PMSF) at a concentration of 1× to 1× buffer A within few minutes of buffer use.

4. Ficoll–sucrose solution: To 80 mL of 1× Buffer A, add 12.5 g sucrose (12.5% w/v final) and 12.5 g Ficoll 400 (12.5% w/v final). Add 1× buffer A to a final volume of 100 mL. Stir or gently rock until sucrose and Ficoll 400 are completely dissolved. Ficoll 400 dissolves slowly. Prepare fresh and use immediately.

5. Deuterium oxide (D_2O; heavy water)–sucrose solution: To 5 mL deuterium oxide (D_2O; heavy water), add: 0.8 g sucrose (8% w/v final) and 1 mL of 10× Buffer A. Add D_2O to 10 mL. Prepare fresh and use immediately.

6. 20% and 50% sucrose solutions in 1× Buffer A: To a 50-mL conical tube containing 25 mL of 1× buffer A (see recipe), add 10 g (20% final) or 25 g (50% final) sucrose. Add 1× buffer A to 50 mL. Stir or gently rock until sucrose is completely dissolved. Prepare fresh and use immediately.

7. Battery-operated pipette filler.

8. Glass–Teflon homogenizers (assorted sizes).

9. Sorvall high-speed centrifuge equipped with SS-34 fixed-angle rotor (or equivalent).

10. Ultracentrifuge with fixed-angle (Beckman 45Ti, Sorvall T-865, or equivalent) and swinging-bucket (Beckman SW-28, Sorvall AH-629, or equivalent) rotors.

11. 25-gauge needle and disposable syringes (assorted sizes).

12. 13-mL thin-walled centrifuge tubes.

13. 2-mm-diameter glass capillary tubes.

14. Two-chamber gradient maker.

15. Peristaltic pump.

16. 5 mg/mL aprotinin, (10,000× stock): To 1 mL H_2O, add 5 mg aprotinin powder (5 mg/mL final) Store up to 6 months at −20 °C.

17. 64 mM benzamidine, (100× stock): To 10 mL H_2O, add 0.1 g benzamidine powder. Pass solution through a 0.22-μm filter. Store up to 6 months at 4 °C.

18. 5 mg/mL leupeptin, (10,000× stock): To 1 mL H_2O, add 5 mg leupeptin powder. Store up to 6 months at −20 °C.

19. 10 mg/mL phenylmethylsulfonyl fluoride (PMSF), (500× stock): To 10 mL of 100% ethanol, add 0.1 g PMSF powder (10 mg/mL final). Store up to 3 months at room temperature.

3   Methods

3.1   Prepare a Crude Homogenate

1. Sacrifice ten Sprague-Dawley rats (150–200 g) by cervical dislocation followed by decapitation or, alternatively, by CO_2 asphyxiation followed by decapitation (*see* **Notes 1** and **2**).

2. Open the rats' skulls with scissors. Remove the brains, transfer them into a beaker and add enough ice-cold 1× buffer A so that they are completely covered (*see* **Note 3**). Keep the beaker on ice (*see* **Note 4**). Once all of the brains have been collected weigh the brains pouring the liquid off and tranfering them to a preweighed beaker (*see* **Note 5**).

3. Once all of the brains have been pooled and weighed, decant and discard the 1× buffer A and transfer the brains to the mortar of a 55-mL glass–Teflon homogenizer.

4. Estimate the volume of homogenization buffer A (containing protease inhibitors) needed: approximately 5 mL of 1× buffer A is needed for every gram of rat brain. Homogenize the tissue using 12 up-and-down strokes at 1500 rpm (*see* **Note 6**). Push the pestle throught the sample rapidly and forcefully.

5. Pool homogenates in a 250-mL graduated cylinder and add ice-cold 1× buffer A to yield a final volume of 10 mL for every gram of rat brain present (e.g., if 20 g of rat brain is present, add ice-cold 1× buffer A to a final volume of 200 mL).

   Reserve a 0.5-mL aliquot of the pooled homogenate (H), but do not discard the rest (*see* **Note 7**).

3.2   Prepare a Microsomal (P2) Fraction by Differential Centrifugation

1. Pour pooled homogenate into six 40-mL centrifuge tubes, balance the tubes, and centrifuge for 20 min at $17,000 \times g$, 4 °C, in a Sorvall SS-34 fixed-angle rotor.

2. Remove the supernatant (S1) from each tube using a pipette filler and pipette, being careful not to disturb the pellet, and combine all supernatants in a 250-mL glass beaker on ice.

3. Resuspend one of the pellets (P1; **step 1**) in 10 mL of 1× buffer A by pipetting up and down using a pipette filler and pipette. Transfer the suspension to the mortar of a 35-mL glass–Teflon homogenizer. Prepare an even suspension using five up-and-down strokes at 1500 rpm.

   Reserve a 0.5-mL aliquot of the suspension (P1) and discard the rest.

4. Transfer S1 (**step 2**) to eight 25-mL centrifuge tubes, balance the tubes, and centrifuge for 60 min at $56,000 \times g$, 4 °C, in a Beckman 45Ti fixed-angle rotor.

5. Remove the supernatant (S2) from each tube using a pipette filler and pipette.

Reserve a 0.5-mL aliquot of the supernatant (S2) from one of the tubes and discard the rest.

6. Resuspend all pellets (P2; **step 4**) using a pipette filler and pipette, in a total of 10 mL of 1× buffer A (*see* **Note 8**).

7. Transfer the resuspended P2 pellets to the mortar of a 35-mL glass–Teflon homogenizer and prepare an even suspension using five up-and-down strokes at 1500 rpm.

8. Transfer the homogenized material to a syringe barrel equipped with a 25-gauge needle. Force the suspension once through the needle by slowly and evenly applying pressure to the plunger and collect in a 50-mL tube.

Reserve a 0.25-mL aliquot of the suspension (P2), but do not discard the rest (*see* **Note 9**).

3.3 Centrifuge Microsomal Fraction on a Ficoll–Sucrose Density Gradient

1. Gently mix by pipetting up and down the P2 suspension (~10 mL) with the same volume (~10 mL) of ice-cold Ficoll–sucrose solution (12.5% w/v). Transfer to a 40-mL centrifuge tube.

2. Centrifuge for 20 min at 43,000 × g, 4 °C, in a Sorvall SS-34 fixed-angle rotor (*see* **Note 10**).

3. Remove the supernatant (sucrose gradient supernatant, or SGs) using a pipette filler and pipette, and transfer to a 100-mL graduated cylinder.

4. Resuspend the pellet (sucrose gradient pellet, or SGp) in 5 mL of 1× buffer A by pipetting up and down. Transfer the suspension to the mortar of a 10-mL glass–Teflon homogenizer and homogenize using five up-and-down strokes at 1500 rpm.

Reserve a 0.5-mL aliquot of the suspension (SGp) and discard the rest.

5. Dilute the SGs from **step 3** (20 mL) by adding 80 mL ice-cold 1× buffer A (1:5 dilution). Transfer to six 25-mL centrifuge tubes.

6. Centrifuge diluted SGs for 60 min at 91,000 × g, 4 °C, in a Beckman 45Ti fixed-angle rotor.

7. Remove and discard all supernatants using a pipette filler and pipette.

8. Combine all pellets and use a pipette filler and pipette to resuspend in a total of 15 mL of 1× buffer A (*see* **Note 11**).

9. Transfer the resuspended material to the mortar of a 30-mL glass–Teflon homogenizer and homogenize using five up-and-down strokes 1500 rpm.

10. Transfer the homogenized material to a syringe barrel equipped with a 25-gauge needle. Force the suspension once through

the needle by slowly and evenly applying pressure to the plunger, and collect in a 40-mL centrifuge tube.

11. Centrifuge for 10 min at 18,700 × g, 4 °C, in a Sorvall SS-34 fixed-angle rotor.

12. Remove the supernatant (still referred to as the SGs) using a pipette filler and pipette, and transfer to a 50-mL graduated cylinder.

Reserve a 0.5-mL aliquot (SGs) and discard the pellet.

3.4 Pellet CCVs Through D₂O–sucrose Cushion

3.4 Pellet CCVs Through D$_2$O–sucrose Cushion

1. Split SGs into 2 aliquots of 7.5 mL and transfer each aliquot to a 15-mL tube. Keep the tube on ice.

2. Pour 2.5 mL D$_2$O–sucrose solution (8% w/v) in a 13-mL thin-walled centrifuge tube designed for a Beckman SW-28 swinging-bucket rotor.

3. Carefully and gently layer on top 7.5 mL of the coated vesicles (SGs) using a Pasteur pipette. Balance the tubes and assemble the buckets and the rotor.

4. Centrifuge samples for 2 h at 116,000 × g, 4 °C, in a Beckman SW-28 swinging-bucket rotor (*see* **Note 12**).

5. Remove and discard the supernatants using a pipette filler and pipette.

6. Combine pellets (cushion-pellet CCVs, or cpCCVs) and resuspend in a total of 1 mL of 1× buffer A plus protease inhibitors.

7. Transfer suspension to the mortar of a 3-mL glass–Teflon homogenizer.

8. Generate an even cpCCV suspension using three up-and-down strokes at 1500 rpm. After homogenization, transfer the suspended material to a single microcentrifuge tube.

Reserve a 100-μL aliquot of the homogenized suspension (cpCCV), but do not discard the rest (*see* **Note 13**).

3.5 Perform Linear Sucrose Gradient Fractionation

3.5 Perform Linear Sucrose Gradient Fractionation

1. Transfer the cpCCV suspension to a syringe barrel equipped with a 25-gauge needle. Pass the suspension once through the needle by slowly and evenly applying pressure to the plunger, and collect suspension in a microcentrifuge tube.

2. Microcentrifuge the cpCCV suspension for 1 min at maximum speed, 4 °C.

3. In a 13-mL thin-walled centrifuge tube designed for a Beckman SW-28 swinging-bucket rotor, prepare a 12-mL linear continuous sucrose gradient from equal volumes of 20% (6 mL) and 50% (6 mL) sucrose in 1× buffer A using a two-chamber gradient maker. Use a peristaltic pump to ensure an even flow rate

from the gradient maker to the centrifuge tube of <2–3 mL/min (*see* **Note 14**).

4. Remove the supernatant (**step 2**) using a Pasteur pipette, and gently layer it onto the top of the linear continuous sucrose gradient. Discard pellet.

5. Centrifuge for 90 min at 145,000 × g, 4 °C, in a Beckman SW-28 swinging-bucket rotor.

6. Collect approximately twelve fractions of 1 mL from the centrifuged gradient in the following way:

    (a) Connect a 2-mm-diameter glass capillary tube to a peristaltic pump using a piece of flexible tubing.

    (b) Gently position the exposed end of the capillary tube so that it rests at the bottom of the tube containing the centrifuged sucrose gradient.

    (c) Activate the peristaltic pump and collect each fraction at a flow rate of ~2–3 mL/min.

7. Use a standard protein assay (e. g., Bradford Protein Assay) to determine which of the isolated fractions yield the highest protein signals (*see* **Note 15**).

8. Pool the 2–3 fractions that yield the highest protein signals and dilute to a total volume of 20 mL in 1× buffer A (1:10 dilution). Transfer to a 25-mL centrifuge tube.

9. Centrifuge the pooled and diluted sample for 60 min at 91,000 × g, 4 °C, in a Beckman 45Ti fixed-angle rotor.

10. Decant and discard supernatant. Resuspend pellet (which contains CCVs) in 0.2 mL of 1× buffer A plus protease inhibitors.

11. Transfer the resuspended CCVs to the mortar of a 1-mL glass–Teflon homogenizer and homogenize using five up-and-down strokes at 1500 rpm.

12. Transfer the suspension to a screw-cap tube, snap-freeze in liquid nitrogen, and store at −70 °C for up to 1 year (*see* **Notes 16** and **17**).

4 Notes

1. All protocols using live animals must first be approved by an Institutional Animal Care and Use Committee (IACUC) and must conform to governmental regulations for laboratory animal care and use. The sacrificing of rats must be supervised or carried out by an experienced animal technician according to specific animal care protocols at the investigator's institution.

2. The protocols described here use fresh tissue, but based on Coomassie Blue staining and immunoblot analysis of subcellular fractions, no obvious differences between fresh and frozen tissue in terms of the protein profiles of the CCVs isolated have been observed. It has been noted that fresh tissue gives a better product for studies of the assembly and disassembly of clathrin coats.

3. A mild acidic pH is essential for the stability of the clathrin coat, so the pH of each buffer should be checked immediately before use. Clathrin coats are also sensitive to protonated amines, and Tris-containing buffers are therefore avoided. In fact, Tris-containing buffers are routinely used to strip clathrin coats following the isolation of CCVs to determine whether a protein of interest is associated with the coat or with the vesicle fraction. For removal of the clathrin coat, CCVs are incubated in 0.3 M Tris–HCl, pH 9.0 (e.g., 50 μL of proteins in a final volume of 1 mL) and rotated for 60 min at 37 °C. The sample is centrifuged for 15 min at $120,000 \times g$ in a Beckman TLA 100.3 rotor with the purified coat fraction remaining in the supernatant, and the uncoated vescicles in the pellet. Resuspend pellet in 1× buffer A (e.g., 50 μL).

4. From this point on, the material is kept on ice throughout the preparation.

5. One rat brain typically weighs 1.5–2.0 g.

6. A large glass–Teflon homogenizer generally has a capacity of 50 mL, so it is likely that at least two homogenizations will be required to process all of the rat brains that have been collected. Do not fill the mortar of the glass–Teflon homogenizer up to the top, but leave space below the rim. Given that during homogenization the investigator does not warm the tissue with his/her hands, the mortar of the glass–Teflon homogenizer should be kept on ice in a small beaker.

7. It is recommended that the purification of CCVs be monitored using clathrin as a marker. Protein aliquots of different subcellular fractions should be therefore retained throughout the protocol that follows for analysis by SDS-PAGE and Coomassie Blue staining and/or immunoblotting. Each fraction that should be included in subsequent analysis is named, and the need to reserve aliquots is indicated. The splitting of suspensions into small aliquots is recommended to eliminate the need for repeated freezing/thawing in the future. All reserved aliquots of different fractions should be collected in screw-cap tubes, snap-frozen in liquid nitrogen and stored at −70 °C for up to 1 year.

8. Ensure that pellets are completely scraped from the walls of the centrifuge tubes. The head of a small Teflon pestle can be used

for this purpose, and the walls of the tubes can be rinsed with the resuspension buffer as well.

9. This first step of purification involves differential centrifugation to generate crude microsomes (P2 fraction), which are enriched for CCVs. These steps reduce contamination by nuclei, mitochondria, and other large organelles, such that the majority of CCV contamination comes from vesiculated endoplasmic reticulum, Golgi bodies, membranes of the endosomal/lysosomal system, and other transport vesicles, including synaptic vesicles.

10. The various forms of density gradient centrifugation employed in this protocol all take advantage of the small size and high density of CCVs to separate these vesicles from contaminating membranes present in the crude microsomal fractions. This centrifugation at $43,000 \times g$ for 20 min leads to pelleting of larger microsomal membranes, while CCVs remain in the supernatant.

11. The head of a small Teflon pestle can be used to scrape the pellets out of their individual centrifuge tubes so that they can be combined.

12. Only dense CCVs are pelleted under these conditions. This step is particularly important when working with brain tissue, since synaptic vesicles, which are abundant in brain, are major contaminating organelles in microsomal fractions. These Ficoll–sucrose and D_2O–sucrose centrifugation steps are effective in separating CCVs from synaptic vesicles.

13. The protocol can be stopped at this point, with the cpCCVs having been collected. The cpCCV suspension can be divided into aliquots to eliminate the need for repente freezing/thawing in the future, transferred to screw-cap tubes, snap-frozen in liquid nitrogen, and stored at −70 °C for up to 1 year. Otherwise, proceed to linear sucrose gradient fractionation.

14. With some practice, linear gradients can be generated reproducibly using a conventional two-chamber gradient maker and a peristaltic pump. Ensure that the solution in the front chamber of the gradient maker (i.e., the chamber with the exit port) is adequately stirred, especially when using high concentrations of sucrose. Gradients should be prepared just before use and placed on ice. Reliable gradients can also be produced by an automatic gradient maker that generates linear gradients following the layering of the densest and lightest solutions in a centrifuge tube.

15. The highest protein signals should be found near the middle of the gradient (around fractions 6–8). The peak protein signal is directly correlated with the peak concentration of CCVs.

16. This CCV suspension can be divided into aliquots before freezing to eliminate the need for repeated freezing/thawing in the future.

17. Using ten rat brains with a total wet weight of 15–20 g, this protocol yields approximately 0.5 mg of CCVs. If the protocol is stopped following the D_2O–sucrose cushion step, the yield is about 1 mg of cushion-pellet CCVs (cpCCVs).

References

1. Kirchhausen T (2014) Clathrin-mediated endocytosis. I: structure and design of CCVs. Cold Spring Harb Perspect Biol. doi: https://doi.org/10.1101/cshperspect.a016725

2. Saheki Y, De Camilli P (2012) Synaptic vesicle endocytosis. Cold Spring Harb Perspect Biol 4(9):a005645

3. Kelly BT, Owen DJ (2011) Endocytic sorting of trans membrane protein cargo. Curr Opin Cell Biol 23:404–412

4. Bonifacino JS, Lippincott-Schwartz J (2003) Coat proteins: shaping membrane transport. Nat Rev Mol Cell Biol 4:409–414

5. Cheng Y, Boll W, Kirchhausen T, Harrison SC, Walz T (2007) Cryo-electron tomography of clathrin-coated vesicles: structural implications for coat assembly. J Mol Biol 365:892–899

6. Daiss JL, Roth TF (1983) Isolation of coated vesicles: comparative studies. Methods Enzymol 98:337–349

7. Pearse BM (1975) Coated vesicles from pig brain: purification and biochemical characterization. J Mol Biol 97:93–98

8. Pearse BM (1982) Coated vesicles from human placenta carry ferritin, transferrin, and immunoglobulin G. Proc Natl Acad Sci U S A 79:451–455

9. Maycox PR, Link E, Reetz A, Morris SA, Jahn R (1992) Clathrin-coated vesicles in nervous tissue are involved primarily in synaptic vesicle recycling. J Cell Biol 118:1379–1388

10. Wasiak S, Legendre-Guillemin V, Puertollano R, Blondeau F, Girard M, de Heuvel E, Boismenu D, Bell AW, Bonifacino JS, McPherson PS (2002) Enthoprotin: a novel clathrin-associated protein identified through subcellular proteomics. J Cell Biol 158:855–862

11. Girard M, Allaire PD, Blondeau F, McPherson PS (2005) Isolation of clathrin-coated vesicles by differential and density gradient centrifugation. Curr Protoc Cell Biol. Chapter 3:Unit 3.13

12. Baudhuin P, Evrard P, Berthet J (1967) Electron microscopic examination of subcellular fractions: I: the preparation of representative samples from suspensions of particles. J Cell Biol 32:181–191

Chapter 2

Preparation of Synaptosomes from Mammalian Brain by Subcellular Fractionation and Gradient Centrifugation

Mirko Messa

Abstract

More than a trillion nerve terminals interconnect neurons in the human brain. These terminals are fundamental for signal transmission and nerve cell communication. Among other techniques, the isolation of nerve terminals [or synaptosomes (Whittaker et al. Biochem J, 90(2):293–303, 1964)] has been fundamental to study the biochemistry and the physiology of the nervous system. This chapter describes the isolation and purification of intact synaptosomes from rodent brain tissue that can be used to further characterize synaptic structure and function and to examine the molecular mechanisms of neurotransmission.

Key words Nerve terminal, Synaptosomes, Homogenization, Percoll™, Density gradient centrifugation

1 Introduction

Almost one hundred billion neurons communicate through nerve terminals in the human brain [1] by generating the signals required for a multitude of functions, such as muscle contraction, hormonal secretion and memory formation. Because of the importance of the synaptic boutons in the nervous system physiology, it became critical to understand the structure and the function of the nerve terminals as well as their regulation [2, 3]. These studies have been made possible by the development of different experimental approaches. One of the most important techniques, from both an historical and experimental point of view, is the isolation of nerve terminals from rodent brain tissue [4–6].

The purification protocol has been described for the first time by Dunkley et al. in 1986 [4] and it has been subjected to multiple modifications since its establishment over the years [6]. Nowadays, the protocols with greater sensitivity and specificity are all based on the homogenization of rat or mouse brain followed by a density

The original version of this chapter was revised. A correction to this chapter can be found at
https://doi.org/10.1007/978-1-4939-8719-1_19

Laura E. Swan (ed.), *Clathrin-Mediated Endocytosis: Methods and Protocols*, Methods in Molecular Biology, vol. 1847,
https://doi.org/10.1007/978-1-4939-8719-1_2, © Springer Science+Business Media, LLC, part of Springer Nature 2018

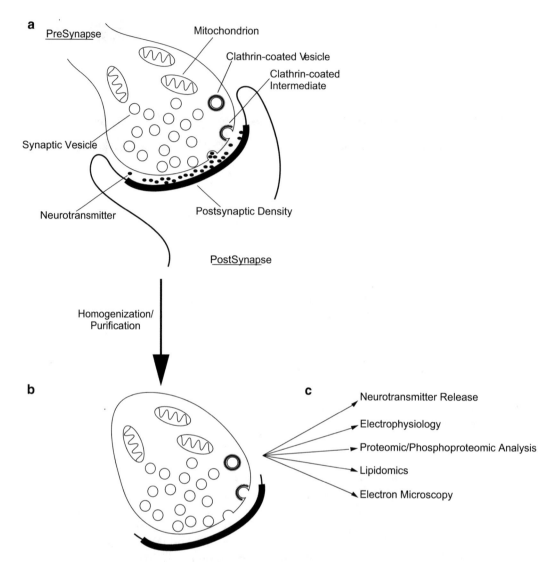

Fig. 1 Composition, preparation, and applications of synaptosomes: Schematic representation of a chemical synapse (**a**) showing the presynaptic terminal containing mitochondria, synaptic vesicles, and endocytic intermediates. The postsynapse faces the presynaptic terminal and a thicker dark line represents the postsynaptic density. Following the homogenization, the presynapse reseals producing synaptosomes that may still have the postsynaptic density attached (**b**). After purification the isolated nerve terminal may be subjected to the different experimental applications (**c**)

gradient centrifugation step on a colloidal suspension of polyvinyl pyrrolidone-coated silica particles known as Percoll™ [6, 7]. In the first step, the synaptic boutons are torn away from the axons and the postsynaptic compartments by using shear force in an isotonic solution at the correct viscosity [8, 9]. Next, the gradient centrifugation removes any debris or contaminants and allows for the purification of reconstituted nerve terminals that are referred as synaptosomes [10]. Morphologically speaking, synaptosomes consist of a resealed presynaptic membrane that encloses the nerve terminal contents (mitochondria, actin cytoskeleton, synaptic vesicles, and cytosolic proteins). However, often the postsynaptic density remains attached to the presynaptic compartment (*see* Fig. 1) and can be detected by electron microscopy [2, 6, 8]. Due to the structure of synaptosomes, they can only be used for acute applications [2]. In fact, they quickly deplete their ATP [11] and do not contain the proper components for new protein synthesis, thus exogenous DNA or siRNA experiments cannot be performed.

Here, I describe a protocol for the purification and isolation of mammalian synaptosomes from rodent brain tissue by using subcellular fractionation coupled with Percoll™ gradient density centrifugation (*see* Fig. 2). The resulting homogenous and intact synaptosomes may be use for biochemical, morphological, and functional studies with aims to characterize the structure of the nerve terminal and the mechanisms of neurotransmission.

2 Materials

– Prepare all solutions using ultrapure water (18 MΩ at 25 °C).

– Before handling any of the reagents or equipment, it is essential to consult the specific Material Safety Data Sheets as well as the Institution's Environmental Health and Safety Office.

2.1 2× Homogenization Buffer (HB) Components

1. 0.5 M EDTA pH 8.0. Add 50 mL of water to a 100 mL glass beaker. Weigh 18.61 g of ethylenediaminetetraacetic acid (EDTA) disodium salt and transfer to the beaker. Adjust the pH to 8.0 with NaOH while stirring. Transfer the solution to a 100 mL graduated cylinder and make up to 100 mL with ultrapure water.

2. 1 M Tris–HCl, pH 7.4. Add 100 mL of water to a 250 mL glass beaker. Weight 24.23 g of Tris and transfer to the beaker. Adjust the pH to 7.4 with HCl while stirring (*see* **Note 1**). Transfer the solution to a 250 mL graduated cylinder and make up to 200 mL with ultrapure water.

3. Protease Inhibitor tablets (cOmplete Tablets, EDTA-free, Roche).

4. 2× Homogenization buffer (HB): 21.9 g of sucrose with 400 μL of 0.5 M EDTA, 2 mL of 1 M Tris pH 7.4 and 70 mL of ultrapure water (*see* **Note 2**) in a 100 mL glass beaker. After the sucrose is completely dissolved, transfer the solution in a 100 mL-graduated cylinder and make up to 200 mL with ultrapure water, (*see* **Note 3**).

2.2 Isotonic Buffer (RB) Components (see Note 4)

1. 5 M NaCl.
2. 3 M KCl.
3. 0.5 M NaHCO$_3$.
4. 0.5 M MgSO$_4$.
5. 0.2 M Phosphate buffer pH 7.4: weigh 35.61 g of sodium phosphate dibasic dihydrate (Na$_2$HPO$_4$-2H$_2$O), transfer to a glass beaker, make up to 1000 mL with ultrapure water and

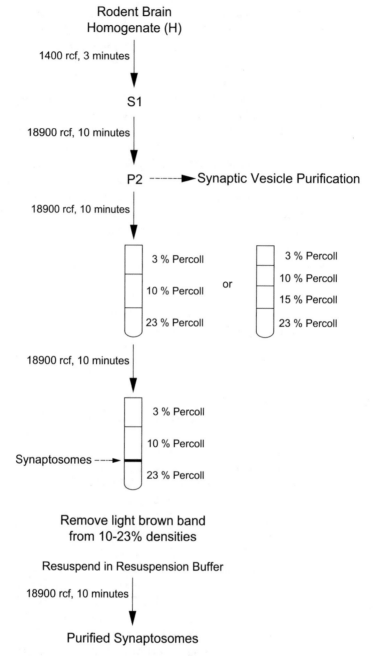

Fig. 2 Schematic representation of the protocol for the purification of synaptosomes from rodent brain tissue

stir until dissolved (*Solution A*). In a second glass beaker transfer 27.6 g of sodium phosphate monobasic monohydrate (NaH$_2$PO$_4$-H$_2$O), make up to 1000 mL with ultrapure water and stir until dissolved (*Solution B*). These two will be your stock solutions; they can be stored at room temperature and used several times to prepare the final phosphate buffer. Mix

40.5 mL of Solution A to 9.5 mL of Solution B to obtain 50 mL of 0.2 M phosphate buffer pH 7.4 (*see* **Note 5**).

6. 10 mM Tris–Hepes buffer pH 7.4: The following buffer is prepared by using a 5 mM unbuffered Tris solution and a 5 mM unbuffered HEPES solution. In a 250 mL glass beaker containing 100 mL water, transfer 0.12 g of Tris, and stir until fully dissolved. Weigh 0.12 g of HEPES, transfer it to a 100 mL-graduated cylinder, make up to 100 mL with ultrapure water and dissolve the HEPES powder. Adjust the pH of the Tris solution by using the unbuffered 0.5 M HEPES until pH 7.4 is reached (*see* **Note 6**). Transfer the final solution in a 250 mL-graduated cylinder and make up to 200 mL with water.

7. Isotonic buffer (IB): Prepare 100 mL of cold Isotonic buffer by mixing 2.8 mL of 5 M NaCl, 167 µL of 3 M KCl, 1 mL of 0.5 M NaHCO$_3$, 260 µL of 0.5 M MgSO$_4$, 250 µL of 0.2 M Phosphate Buffer, and 1 mL of 1 M Tris–Hepes buffer pH 7.4 in 100 mL-graduated cylinder. Make up to 100 mL with water, mix and store at 4 °C.

2.3 Gradient Centrifugation Components

1. Percoll™ (GE Healthcare Life Sciences).
2. Bradford Assay Kit.

2.4 Equipment

1. 15 mL Homogenizer (glass-Teflon™).
2. Motor for the homogenizer pestle.
3. Centrifuge for TA-14 rotor.
4. Centrifuge tubes.
5. Razor blades.
6. 50 mL Falcon tubes.
7. Glass Pasteur pipette with rubber bulb.
8. Gradient Generator (*Optional*, e.g., available from Labconco).
9. TA-14 Rotor from Beckman Coulter or similar (*see* **Note 7**).

3 Methods

The isolation of purified synaptosomes is described below following the schematic diagram protocol showed in Fig. 2.

3.1 Percoll™ Gradient Preparation

1. Mix Percoll™ (*see* **Note 8**), 2× HB and cold water as indicated in Table 1. The quantity showed here are for 4–6 gradients (*see* **Note 9**).
2. Add the Percoll™ solutions in a centrifugation tube (*see* **Note 10**) in the following order (*see* **Note 11**):
   - 2.5 mL of 23% Percoll™ solution.
   - 3 mL of 10% Percoll™ solution.
   - 2.5 mL of 3% Percoll™ solution (*see* **Note 12**).

Table 1
Volume of Percoll™, 2X HB buffer and water necessary for the preparation of 4-6 Percoll™ discontinuous gradients

	Percoll™ (mL)	2× HB (mL)	Water (mL)	Total (mL)
3% Percoll™	0.75	7.5	6.75	15
10% Percoll™	2	10	8	20
23% Percoll™	3.45	7.5	4.05	15

3.2 Rodent Brain Tissue Homogenization

1. Prepare 100 mL of 1× HB by adding and mixing 50 mL of 2× HB with 50 mL of water (*see* **Note 13**).
2. Add Protease Inhibitor tablets (two tablets for 100 mL).
3. Sacrifice one adult rat (or three adult mice, *see* **Note 14**), decapitate and remove the brain tissue. Transfer the brain in an ice-cold glass dish containing 10 mL of 1× HB and chop the tissue in small chunks with a razor blade (*see* **Note 15**).
4. Immediately homogenize the minced brain by using a motor-driven glass-Teflon™ homogenizer containing 10 mL of ice-cold 1× HB using eight up/down stokes at 1000 rpm.
5. Collect the homogenate in a 50 mL Falcon tube, rinse the glass homogenizer with an additional 5 mL of 1× HB in order to have a total of 15 mL rodent brain homogenate (fraction H, *see* **Note 16**).
6. Divide the brain homogenate into two 15 mL centrifuge tubes and centrifuge in a TA-14 rotor at 1400 rcf for 3 min at 4 °C.
7. Save the S1 supernatant (*see* **Note 17**), transfer it to new centrifuge tubes and centrifuge in a TA-14 rotor at 18,900 rcf for 10 min at 4 °C.
8. Remove the S2 supernatant and resuspend the P2 pellet in 1× HB (*see* **Notes 18** and **19**).

3.3 Synaptosome Isolation by Gradient Centrifugation

1. Gently overlay 2 mL of the P2 fraction over a Percoll™ discontinuous gradient (*see* **Note 20**) and centrifuge in a TA-14 rotor at 18900 rcf for 10 min at 4 °C.
2. Collect the band that has formed between the 23% and 10% Percoll™ interface (*see* Fig. 3) and transfer it to a clean centrifuge tube (*see* **Note 21**).
3. Add 20 mL of IB to the tube and centrifuge in a TA-14 rotor at 18,900 rcf for 10 min at 4 °C (*see* **Note 22**).
4. Remove the supernatant and resuspend the pellet (washed, purified synaptosomes) with 2–3 mL of IB.
5. Determine the concentration of the isolated purified synaptosomes by using the Bradford assay (*see* **Note 23**).

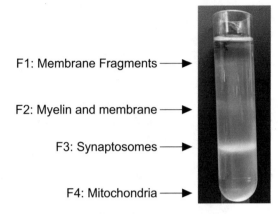

F1: Membrane Fragments ⟶

F2: Myelin and membrane ⟶

F3: Synaptosomes ⟶

F4: Mitochondria ⟶

Fig. 3 Results of the density gradient centrifugation (of a three-layer Percoll™ gradient). *F1*: membrane fragments, *F2*: myelin and membrane, *F3*: Synaptosomes and *F4*: mitochondria

3.4 Use of Purified Synaptosomes

The purified synaptosomes may be used for several applications. The intact nerve terminals can be easily manipulated by pharmacological tools and allow scientists to use them for functional studies such as neurotransmitter release or for biochemical studies of the molecular machinery and posttranslational modifications of proteins involved in neurotransmission [12–15]. Fluorescence [13, 16, 17] and electron microscopy [6, 18, 19] imaging may be used as well to address fundamental questions not only in term of function, but also of architecture of the mammalian synaptic terminal.

4 Notes

1. Warm water (37 °C) helps to easily dissolve Tris. Adjust the pH after the solution has been brought to room temperature.
2. Dissolve the sucrose in prechilled water.
3. Stir and store the solution at 4 °C.
4. The recipe here reported is an example of an isotonic solution that may be used to resuspend the synaptosomes after the gradient centrifugation. Depending on the application other solutions may be used, check the following references for more information [6, 13, 20, 21].
5. Verify that the pH is 7.4 just before using the final solution.
6. The final molarity of the Tris/HEPES solution corresponds to the buffer capacity of the individual Tris and HEPES buffers.
7. As an alternative, the SS-34 rotor from Thermo Scientific may be used.
8. Percoll™ is stored at 4 °C. Before preparing the different density solutions filter 10 mL of Percoll™ using a 0.45 µm filter.

The remaining filtered Percoll™ can be stored at 4 °C and should be used within 1 month.

9. Since the gradients may break and cause the different densities to mix, prepare them in excess to have four good three-layer Percoll™ gradients for the purification step.

10. Add the different solutions by using a glass Pasteur pipette and a rubber bulb or a gradient-maker. The use of a gradient maker may generate a more homogenous gradient. For a specific protocol, follow the gradient manufacturer's instructions.

11. Prepare the gradient at 4 °C. Carefully store the gradients at 4 °C for 24 h maximum.

12. The described protocol uses a three-layer Percoll™ gradient. For some specific applications, it may be useful to add a 15% Percoll™ layer to remove additional membrane contaminations [6]. At the same time, a four-layer gradient will reduce the final synaptosomal yield.

13. Use previously chilled water and store at 4 °C.

14. Carefully follow your institutional guidelines for the care and use of laboratory animals.

15. Although it has been reported that synaptosomes may be purified from frozen tissue [22] it is preferable to always use fresh brain tissue to obtain a more homogenous preparation.

16. From here on, save an aliquot of each fraction you generate to check the quality of your purification by Western blot.

17. Discard P1 pellet, it mostly contains white matter, blood, and other debris from the brain tissue homogenization.

18. P2 fraction may be used to purify synaptic vesicles from rodent brain tissue [23–25]

19. Use 8 mL for four gradients.

20. Add the P2 fraction over the gradient by using a glass Pasteur pipette.

21. To avoid excessive contamination from the other bands in the gradient, remove the synaptosome fraction by using a glass Pasteur pipette.

22. This step is fundamental; wash the isolated purified synaptosomes from any residual Percoll™ and remove any possible debris from other bands in the gradients.

23. A protein concentration of 2–3 mg/mL is expected from one rat brain after resuspension of the washed synaptosome fraction in 2–3 mL of IB.

References

1. Azevedo FAC, Carvalho LRB, Grinberg LT et al (2009) Equal numbers of neuronal and nonneuronal cells make the human brain an isometrically scaled-up primate brain. J Comp Neurol 513(5):532–541. https://doi.org/10.1002/cne.21974

2. Whittaker VP (1993) Thirty years of synaptosome research. J Neurocytol 22(9):735–742. https://doi.org/10.1007/BF01181319

3. Thorne B, Wonnacott S, Dunkley PR (1991) Isolation of hippocampal synaptosomes on Percoll gradients: cholinergic markers and ligand binding sites. J Neurochem 56(2):479–484. https://doi.org/10.1111/j.1471-4159.1991.tb08175.x

4. Dunkley PR, Jarvie PE, Heath JW, Kidd GJ, Rostas JA (1986) A rapid method for isolation of synaptosomes on Percoll gradients. Brain Res 372(1):115–129

5. Dunkley PR, Heath JW, Harrison SM, Jarvie PE, Glenfield PJ, Rostas JA (1988) A rapid Percoll gradient procedure for isolation of synaptosomes directly from an S1 fraction: homogeneity and morphology of subcellular fractions. Brain Res 441(1–2):59–71

6. Dunkley PR, Jarvie PE, Robinson PJ (2008) A rapid Percoll gradient procedure for preparation of synaptosomes. Nat Protoc 3(11):1718–1728. https://doi.org/10.1038/nprot.2008.171

7. Matsumoto I, Combs MR, Jones DJ (1992) Characterization of 5-hydroxytryptamine1B receptors in rat spinal cord via [125I]iodocyanopindolol binding and inhibition of [3H]-5-hydroxytryptamine release. J Pharmacol Exp Ther 260(2):614–626

8. Gray EG, Whittaker VP (1962) The isolation of nerve endings from brain: an electron-microscopic study of cell fragments derived by homogenization and centrifugation. J Anat 96(Pt 1):79–88. https://doi.org/10.1111/(ISSN)1469-7580

9. De Robertis E, Pellegrino de Iraldi A, Rodriguez de Lores Arnaiz G, Salganicoff L (1962) Cholinergic and non-cholinergic nerve endings in rat brain-i isolation and subcellular distribution of acetylcholine and acetylcholinesterase. J Neurochem:1–19

10. Whittaker VP, Michaelson IA, Kirkland RJ (1964) The separation of synaptic vesicles from nerve-ending particles ("synaptosomes"). Biochem J 90(2):293–303

11. Dunkley PR, Robinson PJ (1986) Depolarization-dependent protein phosphorylation in synaptosomes: mechanisms and significance. Prog Brain Res 69:273–293

12. Musante V, Summa M, Cunha RA, Raiteri M, Pittaluga A (2011) Pre-synaptic glycine GlyT1 transporter--NMDA receptor interaction: relevance to NMDA autoreceptor activation in the presence of Mg2+ ions. J Neurochem 117(3):516–527. https://doi.org/10.1111/j.1471-4159.2011.07223.x

13. Musante V, Neri E, Feligioni M et al (2008) Presynaptic mGlu1 and mGlu5 autoreceptors facilitate glutamate exocytosis from mouse cortical nerve endings. Neuropharmacology 55(4):474–482. https://doi.org/10.1016/j.neuropharm.2008.06.056

14. Jovanovic JN, Sihra TS, Nairn AC, Hemmings HC, Greengard P, Czernik AJ (2001) Opposing changes in phosphorylation of specific sites in synapsin I during Ca2+−dependent glutamate release in isolated nerve terminals. J Neurosci 21(20):7944–7953

15. Vargas KJ, Makani S, Davis T, Westphal CH, Castillo PE, Chandra SS (2014) Synucleins regulate the kinetics of synaptic vesicle endocytosis. J Neurosci 34(28):9364–9376. https://doi.org/10.1523/JNEUROSCI.4787-13.2014

16. Serulle Y, Sugimori M, Llinás RR (2007) Imaging synaptosomal calcium concentration microdomains and vesicle fusion by using total internal reflection fluorescent microscopy. Proc Natl Acad Sci 104(5):1697–1702. https://doi.org/10.1073/pnas.0610741104

17. Choi SW, Gerencser AA, Lee DW et al (2011) Intrinsic bioenergetic properties and stress sensitivity of dopaminergic synaptosomes. J Neurosci 31(12):4524–4534. https://doi.org/10.1523/JNEUROSCI.5817-10.2011

18. Villasana LE, Klann E, Tejada-Simon MV (2006) Rapid isolation of synaptoneurosomes and postsynaptic densities from adult mouse hippocampus. J Neurosci Methods 158(1):30–36. https://doi.org/10.1016/j.jneumeth.2006.05.008

19. De Camilli P, Cameron R, Greengard P (1983) Synapsin I (protein I), a nerve terminal-specific phosphoprotein. I. Its general distribution in synapses of the central and peripheral nervous system demonstrated by immunofluorescence in frozen and plastic sections. J Cell Biol 96(5):1337–1354

20. Cousin MA, Robinson PJ (2000) Two mechanisms of synaptic vesicle recycling in rat brain nerve terminals. J Neurochem 75(4):1645–1653

21. Robinson PJ, Gehlert DR, Sanna E, Hanbauer I (1989) Two fractions enriched for striatal synaptosomes isolated by Percoll gradient centrifugation: synaptosome morphology, dopamine and serotonin receptor distribution,

and adenylate cyclase activity. Neurochem Int 15(3):339–348

22. Gleitz J, Beile A, Wilffert B, Tegtmeier F (1993) Cryopreservation of freshly isolated synaptosomes prepared from the cerebral cortex of rats. J Neurosci Methods 47(3):191–197. https://doi.org/10.1016/0165-0270(93)90081-2

23. Huttner WB, Schiebler W, Greengard P, De Camilli P (1983) Synapsin I (protein I), a nerve terminal-specific phosphoprotein. III. Its association with synaptic vesicles studied in a highly purified synaptic vesicle preparation. J Cell Biol 96(5):1374–1388. https://doi.org/10.1083/jcb.96.5.1374

24. Messa M, Congia S, Defranchi E et al (2010) Tyrosine phosphorylation of synapsin I by Src regulates synaptic-vesicle trafficking. J Cell Sci 123(Pt 13):2256–2265. https://doi.org/10.1242/jcs.068445

25. Onofri F, Messa M, Matafora V et al (2007) Synapsin phosphorylation by Src tyrosine kinase enhances Src activity in synaptic vesicles. J Biol Chem 282(21):15754–15767. https://doi.org/10.1074/jbc.M701051200

Chapter 3

Probing Endocytosis During the Cell Cycle with Minimal Experimental Perturbation

António J. M. Santos and Emmanuel Boucrot

Abstract

Endocytosis mediates the cellular uptake of nutrients, modulates signaling by regulating levels of cell surface receptors, and is usurped by pathogens during infection. Endocytosis activity is known to vary during the cell cycle, in particular during mitosis. Importantly, different experimental conditions can lead to opposite results and conclusions, thereby emphasizing the need for a careful design of protocols. For example, experiments using serum-starvation, ice-cold steps or using mitotic arrest produced by chemicals widely used to synchronize cells (nocodazole, RO-3306, or S-trityl-L-cysteine) induce a blockage of clathrin-mediated endocytosis during mitosis not observed in unperturbed, dividing cells. In addition, perturbations produced by mRNA interference or dominant-negative mutant overexpression affect endocytosis long before cells are being assayed. Here, we describe simple experimental procedures to assay endocytosis along the cell cycle with minimal perturbations.

Key words Endocytosis, Clathrin-mediated endocytosis, Cell cycle, Mitosis, Cell division, Cell synchronization, Serum starvation, Preincubation on ice, Nocodazole, RO-0336, S-Trityl-L-cysteine, siRNA, Dynasore, Pitstop, Knock-sideways

1 Introduction

Endocytosis is essential to all eukaryotic cells for nutrient uptake and for controlling the levels of receptors, channels, and transporters at their surface. There are several pathways of endocytosis defined by the receptors that use them, by their distinct morphological features or by their requirement of key cytosolic components [1]. Clathrin-mediated endocytosis (hereafter, CME) is the best characterized portal of entry into cells [2, 3]. By contrast, clathrin-independent endocytosis has been less well understood so far [1, 4, 5].

Clathrin-coated pits are formed constitutively at the surface of cells upon polymerization of clathrin triskelia into lattices wrapping around the invaginating membrane. Receptors destined to

The original version of this chapter was revised. A correction to this chapter can be found at
https://doi.org/10.1007/978-1-4939-8719-1_19

Laura E. Swan (ed.), *Clathrin-Mediated Endocytosis: Methods and Protocols*, Methods in Molecular Biology, vol. 1847,
https://doi.org/10.1007/978-1-4939-8719-1_3, © Springer Science+Business Media, LLC, part of Springer Nature 2018

be internalized (also called "cargo" molecules) are sorted into clathrin-coated pits by various adaptor proteins as clathrin itself cannot bind to receptors or to membrane [3, 6]. The tetrameric adaptor complex AP2 is central to CME, forming an interaction hub between receptors, clathrin, lipids, and associated adaptor proteins [3, 6]. Some cargoes, in particular transferrin and low-density lipoprotein (LDL) receptors, rely almost exclusively (>70–80%) on CME to enter cells and their rates of uptake are widely used as readout of the activity of the pathway [3]. Another assay commonly used to measure CME is recording clathrin-coated pit and vesicle formation using live-cell imaging fluorescent microscopy [7, 8].

Inhibition of CME is best achieved in mammalian cells by gene silencing (RNA interference (RNAi) or knockout) of clathrin heavy chain or AP2 subunits [9, 10] or by sequestering them into the cytoplasm upon overexpression of high affinity binding protein domains, such as AP180 C-terminus (aa530-915) or Eps15 Δ95/295 mutants [11, 12]. CME is also inhibited upon overexpression of dynamin function-defective mutants such as K44A [13] or T65A [14], although dynamin functions in other clathrin-independent endocytic pathways.

However, it is important to be mindful of the timeframe of the perturbations considering the biological process being studied. Endocytosis happens within minutes but gene silencing or mutant overexpression required several days to be efficient, thus CME will be measured well after the perturbation has begun to occur. Thus, any phenotypes observed during the cell cycle stage studied might result from the compounded effect of the perturbations of endocytosis for many hours—or days—prior to the assay instead of a direct effect on CME.

Typically, RNAi requires at least 3 days for the levels of the protein of interest to decrease by 80–90% of control. RNA interference blocks translation of targeted proteins by degrading their mRNA, thereby blocking production of new protein. But the existing, original, pool of the protein of interest is still present in the cells. Its levels are gradually decreased by proteasome or lysosomal degradation (normal recycling of proteins) and by dilution upon cell division. Thus, assuming an optimal inhibition of mRNA translation of a very long-lived protein (such as clathrin heavy chain, which turnover rate is as long as the cell cycle), it takes theoretically three cell divisions to dilute protein levels down to ~12.5% of control levels (that is, ~87.5% depletion) (see Fig. 1). But by the time one assays the samples on day 3, cells will have divided twice (during day 1 and 2) with ~50% and ~25% of initial protein levels, respectively. This decrease will induce defects that cumulated and that will contribute greatly to any phenotype observed during the assay.

Therefore it is desirable to minimize indirect adverse effects of the experimental protocols. Favorable options are to time the

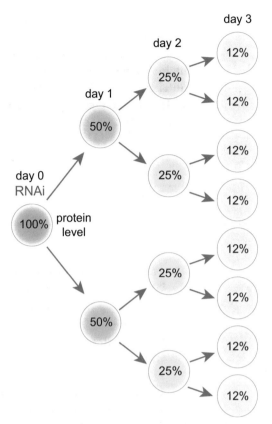

RNAi - protein dilution upon cell division

Fig. 1 Protein dilution by cell division during RNA interference. Assuming an optimal inhibition of mRNA translation of a very long-lived protein, it takes theoretically three cell divisions to dilute the existing protein levels down to ~12.5% of control levels (that is, ~87.5% depletion)

overexpression of dominant-negative mutants that its onset coincides with the cell cycle stage studied (as described below) or to use rapid (minutes) inactivation methods using small compound inhibitors such as Dynasore, Pitstop 2 [15, 16], or acute protein sequestration, or "knock-sideways" [17].

Although widely used in many studies, some experimental conditions such as cold preincubation or serum starvation prior to ligand uptake affect their endocytosis [18]. Historically used to artificially enhance endocytosis by upregulating receptor levels on the cell surface (serum starvation) or to synchronize entry (preincubation of cells with ligands on ice), these steps introduce bias into the experiments by affecting cell physiology and membrane biophysical properties. Serum starvation, usually performed for 16 h (overnight) prior to endocytic assays, activates autophagy within a few hours [19], thereby changing the endolysosomal system of the cells to be assayed. Preincubation of cells with ligands on ice affects plasma membrane fluidity (likely inducing phase transition), and favors ligand cross-linking, which will impact endocytosis following warming up back to 37 °C. These experimental conditions were also reported to upregulate clathrin-independent pathways, and thus affecting cell physiology. In addition, chemical synchronization used to stall cells at various stages of the cell cycle (serum starvation, double thymidine block, nocodazole, RO-3306 or S-trityl-L-cysteine) also affect endocytosis [20–22]. This chapter outlines methods that we used to measure endocytosis along the cell cycle with minimal experimental perturbations.

2   Materials

1. Primary, normal cells or cell lines cultured in their appropriate complete growth medium (*see* **Note 1**).

2. Treatments being investigated (inhibitor drugs, siRNA, plasmids, etc.). Here, we provide guidance for the use of chemical CME inhibitors (Pitstop2 (25 mM stock in DMSO), Dynasore (80 mM stock in DMSO), or Dyngo-4a (10 mM stock in DMSO)) and overexpressed inhibitory proteins (AP180 C-terminus (aa530-915) and Eps15 Δ95/295, and function-defective mutants such as Dynamin K44A or T65A).

3. Phosphate-buffered saline (PBS) buffer: 138 mM NaCl, 2.7 mM KCl, 10 mM Na_2HPO_4, 1.76 mM KH_2PO_4 equilibrated at pH 7.4.

4. EDTA-based cell detachment solution: 0.4 mM EDTA in PBS without Ca^{2+} and Mg^{2+}.

5. Transfection reagent as appropriate for the cell type used (e.g., Lipofectamine 2000, Thermo Scientific).

6. Fluorescently labeled ligands to monitor endocytosis (e.g., Alexa 488-Transferrin, Thermo Scientific).

7. Stripping buffer 1 (pH 5.5): 150 mM NaCl, 20 mM HEPES, 5 mM KCl, 1 mM $CaCl_2$, 1 mM $MgCl_2$, adjusted to pH 5.5.

8. Stripping buffer 2 (pH 2.5): 150 mM NaCl, 0.2 M acetic acid, 5 mM KCl, 1 mM $CaCl_2$, 1 mM $MgCl_2$, adjusted to pH 2.5.

9. Paraformaldehyde (PFA), diluted to 3.7% in PBS.

10. 50 mM NH_4Cl in PBS

11. Ligand/antibody uptake assay medium: α-MEM without phenol red supplemented with 20 mM Hepes, pH 7.2.

12. Imaging medium: α-MEM without phenol red supplemented with 20 mM Hepes, pH 7.2 and 5% FBS)

3   Methods

3.1   Optimization of Cell Cycle Without Drug Synchronization

Upon confluency, cells terminate their current cell cycle and naturally pause in early G1 [23]. After several hours following contact, normal (i.e., untransformed) cells exit the cell cycle to become quiescent (known as the G0 stage of the cell cycle) [24]. Transformed cells having lost contact inhibition stall first in their cell cycle in early G1 upon reaching confluency (*see* Fig. 2a, b), but after several hours, can extravasate to continue into another cell cycle. Thus, one can take advantage of the initial, contact-induced, stalling of both normal and transformed cells in early G1 to time cell cultures. When cells are passaged few hours after reaching confluency, a greater proportion will resume their cell cycle at the same stage (early G1) and an increased proportion (up to 50% increase) of cells will concomitantly reach S, G2, and M phases (*see* Fig. 2c) than in cultures growing exponentially (and thus asynchronously) [21, 23, 25].

This simple procedure takes advantage of the natural pause following cell–cell contact (*see* Fig. 2c), and provides an opportunity to naturally enrich cell cultures at each stage of the cell cycle. This requires that the experimenter knows the cell cycle length of cell type used for the assays. This is determined using classic cell doubling time measurements and flow cytometry cell cycle analysis (*see* Fig. 2a). In our experience, it varies between 14 and 28 h between cell types.

1. Grow cells to become fully confluent for ~24 h (*see* **Note 2**).

2. Detach cells using EDTA-based cell detachment solution (*see* **Note 3**).

3. Seed cells at ~70% confluence in their appropriate complete culture medium (*see* **Note 4**).

G1 phase: After ~30% of the cell cycle length (~8 h for a 24 h cell cycle, Fig. 2b, c), over 90% of cells should be in G1, as compared to 50–70% in asynchronized populations (*see* **Note 5**).

S phase: After ~70% of the cell cycle length (~17 h for a 24 h cell cycle, Fig. 2b, c), the majority of cells should be in S, as compared to 15–20% in asynchronized populations (*see* **Note 6**).

G2 phase: After ~90% of the cell cycle length (~21 h for a 24 h cell cycle, Fig. 2b, c), the majority of cells should be in G2, as compared to 15–20% in asynchronized populations (*see* **Note 7**).

M phase: after ~100% of the cell cycle length (~24 h for a 24 h cell cycle, Fig. 2b, c), there is a wave of cells naturally undergoing mitosis lasting 2–3 h. During this time window, the mitotic index (% of cell undoing mitosis) can be up to 30%, as compared to 1–5% in asynchronized populations (*see* **Note 8**).

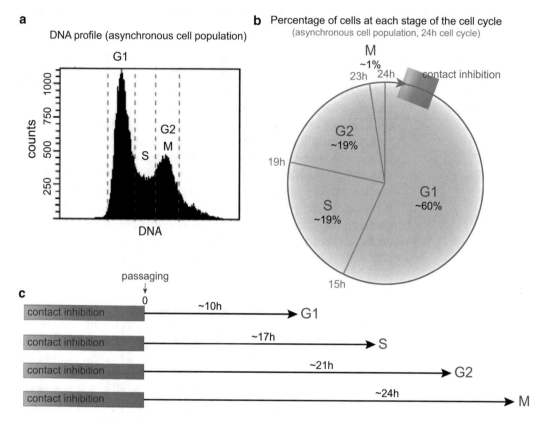

Fig. 2 Optimization of cell cycle without drug synchronization. (**a**) DNA profile of an asynchronous RPE1 cell population. DNA was labeled using Hoechst 33342. "2n" DNA peak correspond to cells in G1, "4n" DNA peak corresponds to cells in G2 or M and cells having between "2n" and "4n" DNA are ongoing S phase. (**b**) Percentage of cells at each stage of the cell cycle calculated from a DNA profile as in (**a**) for a cell type having a 24 h cell cycle (measured population doubling time). Natural contact inhibition period (early G1) in confluent cells is indicated in red. (**c**) Cell plating protocol to increase the percentage of cells in G1, S, G2, or M, respectively

3.2 Inhibiting CME at Specific Cell Cycle Stages Using Mutant Protein Overexpression

Clathrin-mediated endocytosis is inhibited by overexpressing protein fragments that sequester clathrin or AP2 in the cytoplasm (AP180 C-terminus (aa530-915) and Eps15 Δ95/295, respectively [11, 12] or function-defective mutants such as Dynamin K44A [13] or T65A [14].

1. On test samples, determine the minimal time after transfection after which the construct is expressed. Fast-folding proteins are expressed as soon as ~8 h after transfection, whereas slow maturing ones require over 24 h (*see* **Note 9**).

2. Passage and transfect cells for the appropriate number of hours prior to the assay (*see* Fig. 3a). For example, using a protein expressing 10 h after transfection and a cell line having a 24 h cell cycle, time the passaging and transfection as follow:

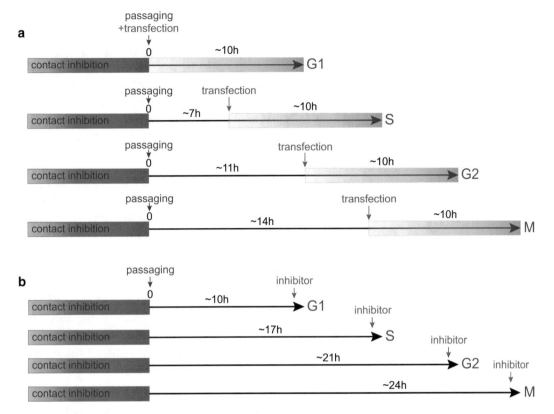

Fig. 3 Inhibiting CME at specific cell cycle stages using mutant protein overexpression or small inhibitors. (**a**) Cell plating and transfection protocol to inhibit CME in cells in G1, S, G2, or M, respectively. (**b**) Cell plating and small inhibitor addition protocol to inhibit CME in cells in G1, S, G2, or M, respectively

G1 phase: Transfect cells just after plating (also called "reverse transfection," when the cells are still suspended). After ~10 h, most transfected cells will be in G1 (*see* Fig. 3a).

S phase: Plate the cells as in Subheading 3.1. Transfect cells ~7 h after plating. After ~10 h (thus ~17 h after cell plating) most transfected cells will be in S (*see* Fig. 3a).

G2 phase: Plate the cells as in Subheading 3.1. Transfect cells ~11 h after plating. After ~10 h (thus ~21 h after cell plating) most transfected cells will be in G2 (*see* Fig. 3a).

M phase: Plate the cells as in Subheading 3.1. Transfect cells ~14 h after plating. After ~10 h (thus ~24 h after cell plating) most transfected cells will be in mitosis (*see* Fig. 3a). For a transfection efficiency of ~30%, and a mitotic index of ~30%, one can expect ~10% of all cells to be transfected and undergoing mitosis at the time of the assay (*see* **Note 10**).

3.3 Inhibiting CME at Specific Cell Cycle Stages Using Small Molecule Inhibitors

Using the cell seeding procedure detailed in Subheading 3.1, treat cell cultures in G1, S, G2, or M phase with the appropriate small inhibitors for the appropriate time before the assays (*see* Fig. 3b).

Dynamin inhibitors: dilute the inhibitor to the appropriate concentration into culture medium (Dynasore: 80 μM, Dyngo-4a: 10 μM [15, 26]. Replace culture medium with medium containing the inhibitor and incubate at 37 °C for the appropriated time (a minimum of 10 min and maximum of 60 min is recommended) (*see* **Note 11**).

Clathrin inhibitors: dilute the inhibitor to the appropriate concentration into culture medium (Pitstop 2: 50 μM [16]). Replace the culture medium with the medium containing the inhibitor and incubate at 37 °C for the appropriate time (a minimum of 10 min and maximum of 60 min is recommended) (*see* **Note 12**).

3.4 Fluorescently Tagged Ligand Uptake Endocytic Assays in Unperturbed Cells

Although widely used in the literature, cold preincubation or serum starvation prior to ligand uptake both affect their endocytosis [18]. To minimize experimental perturbations, the following assays use direct incubation of ligands or antibodies.

Cells seeded on glass coverslips or glass-bottom microplates should be prepared according to Subheadings 3.1 and 3.2, as required, prior to the assay.

1. Prewarm (37 °C for 1 h) ligand uptake assay medium (*see* **Note 13**).

2. Dilute fluorescently tagged ligand to the desired concentration (e.g., 50 μg/mL Tf-Alexa 488) in warmed ligand uptake assay medium.

3. Take cells out of the incubator, swiftly aspirate culture medium and replace it with prewarmed assay medium containing ligand, return the cells to the 37 °C incubator, and start a timer (*see* **Note 14**).

4. After desired time has elapsed, remove the cells from the incubator, aspirate assay medium, add ice-cold PBS and place the cells on a water-ice tray (*see* **Note 15**).

5. Wash the cells another time with ice-cold PBS, and incubate them on ice with stripping buffer 1 for 5 min (*see* **Note 16**).

6. Remove stripping buffer 1 and replace with fresh ice-cold stripping buffer 1 and incubate for another 5 min on ice.

7. Wash the cells three times with ice-cold PBS (*see* **Note 17**).

8. Add fixative solution (3.7% PFA in PBS) and incubate on ice for 20 min.

9. Wash the fixed cells three times with PBS containing 50 mM NH_4Cl (*see* **Note 18**).

10. Immunostain cells as required (*see* **Note 19**).

3.5 Antibody Uptake Feeding Assays

Cells seeded on glass coverslips or glass-bottom microplates should be prepared according to Subheadings 3.1 and 3.2, as required, prior to the assay.

1. Prewarm (37 °C for 1 h) antibody uptake assay medium (*see* **Note 13**).

2. Dilute antibody to the desired concentration (e.g., 5 µg/mL for an antibody against the ectodomain of EGFR).

3. Take cells out of the incubator, swiftly aspirate culture medium and replace it with prewarmed assay medium containing antibodies, return the cells to the 37 °C incubator and start a timer (*see* **Note 14**).

4. After the desired time has elapsed, remove the cells from the incubator, aspirate assay medium, add ice-cold PBS and place the cells on a water-ice tray (*see* **Note 15**).

5. Wash the cells another time with ice-cold PBS, and incubate them on ice with stripping buffer 2 for 5 min (*see* **Note 20**).

6. Remove stripping buffer 2 and replace it with fresh ice-cold stripping buffer 2 and incubate for another 5 min.

7. Wash the cells three times with ice-cold PBS (*see* **Note 17**).

8. Add fixative solution (3.7% PFA in PBS) and incubate on ice for 20 min.

9. Wash the fixed cells three times with PBS containing 50 mM NH_4Cl (*see* **Note 18**).

10. Immunostain cells as required (*see* **Note 21**).

3.6 Spatiotemporal Measurements of Clathrin-Coated Pit and Vesicle Formation

Gene-edited cells or cells stably expressing EGFP-tagged AP2 (σ2-EGFP [27] or EGFP-α2 [28]) or clathrin (EGFP-CLCa [29]) should be seeded on glass-bottom dishes or microplates and prepared according to Subheadings 3.1 and 3.2, as required, prior to the assay.

1. Prewarm (37 °C for 1 h) imaging medium (*see* **Note 22**).

2. Replace culture medium with imaging medium.

3. Image clathrin-coated pit and vesicle formation using a spinning-disc confocal [27], a total-internal reflection (TIRF) [30], or a light-sheet [20] microscope equipped with a 37 °C a 5% CO_2 incubation chamber (*see* **Note 23**).

4. Quantify clathrin-coated pit and vesicle formation using automated tracking software (e.g., [31]).

4 Notes

1. A wide vamriety of cells may be used. The methods described here use human normal, diploid hTERT-RPE1 cells, but are easily adapted for other cell types.

2. About 24 h after cell confluency is required to induce contact inhibition of growth [23, 24].

3. The timings presented in 3.1 were set up using cells detached upon calcium chelation (EDTA-based cell detachment). Detaching cells with trypsin-based solutions will likely require longer times as it shaves cells from a significant proportion of their cell surface integrins and thus slows reattachment onto the substratum.

4. We found that seeding cells at this confluency was required for a reentry into the cell cycle following the timeframes presented. Seeding cells at lower density decreases the proportion of cells at each stage of the cell cycle after the respective timeframes presented in Subheading 3.1.

5. The proportion of cells actually in G1 should be confirmed using flow cytometry (proportion of "2n" cells).

6. The proportion of cells actually in S should be confirmed using flow cytometry.

7. The proportion of cells actually in G2 should be confirmed using flow cytometry.

8. Mitotic substages (prophase, metaphase, anaphase, telophase, and cytokinesis) can be easily distinguished using bright-field or phase-contrast microscopy.

9. Choosing the minimum time following transfection after which the mutant protein is expressed is crucial to minimize indirect effects of blocking endocytosis for long periods of time before the cell cycle of interest.

10. This procedure generates a low number of transfected cells undergoing mitosis (~10%) but single-cell approaches (microscopy or flow cytometry) can easily identify them.

11. Other dynamin inhibitors such as Dynole 34-2, OctMAB, MitMAB, Rhodadyn C10, or Iminodyn-22 [32–34] can be used as well.

12. Other chemical inhibitions of CME such as hypertonic shock (0.45 M sucrose), potassium depletion, monodansylcadaverine, chlorpromazine, or phenylarsine oxide [35–39], even though widely used in the literature are not recommended as they affect clathrin-independent endocytosis as well [40].

13. Prewarming ligand/antibody uptake medium is required to maintain the temperature of the cells as close to 37 °C as possible throughout the assay.

14. Incubations times can vary from 1 min to 60 min, as required. To measure endocytic rates, several lengths of incubations must be measured.

15. The use of cold temperatures at the end of the assay is appropriate as the cells will be soon fixed. Adding ice-cold PBS stops any trafficking within seconds. Using ice + water mix to cool the plates/dishes during the subsequent steps insure a better temperature exchange and cooling than using ice only.

16. Washes with a mildly acidic stripping buffer ("Stripping buffer 1," pH 5.5) remove cell surface ligands that were not internalized after incubation.

17. Washes with PBS are required for raising back the pH to ~7.4 after washes with stripping buffers.

18. Washes with 50 mM NH_4Cl in PBS are required to inactivate any residual reactive PFA.

19. Counterstaining of protein(s) of interest, DNA (using DAPI or DRAQ5) and/or actin cytoskeleton (phalloidin) using other fluorophores than those coupled on the internalized ligand can be considered.

20. Washes with an acidic stripping buffer ("Stripping buffer 2," pH 2.5) remove cell surface antibodies that were not internalized after incubation. Note that cell surface stripping of antibodies requires lower pH than for endocytic ligands.

21. Fluorescently labeled secondary antibodies toward the specie of internalized antibodies must be included in the immunostaining procedure (e.g., Alexa 488 coupled goat anti-rabbit secondary antibodies to label rabbit anti-EGFR antibodies internalized during the feeding assay).

22. Prewarming imaging medium is required to maintain the temperature of the cells as close to 37 °C as possible throughout the assay and to maintain clathrin-coated pits and vesicles dynamics.

23. Clathrin-coated pit and vesicle formation in mitotic cells can be imaged on the bottom surface (contacting the glass coverslip) using TIRF microscopy, both on the top or bottom surfaces using spinning-disk confocal microcopy or throughout entire cells using light-sheet super-resolution microscopy [20, 27, 30].

Acknowledgments

A.J.M.S. was supported by Fundação para a Ciência e Tecnologia (SFRH/BD/33545/2008), and E.B. was a Biotechnology and Biological Sciences Research Council (BBSRC) David Phillips Research Fellow.

References

1. Doherty GJ, McMahon HT (2009) Mechanisms of endocytosis. Annu Rev Biochem 78:857–902

2. Kirchhausen T, Owen D, Harrison SC (2014) Molecular structure, function, and dynamics of clathrin-mediated membrane traffic. Cold Spring Harb Perspect Biol 6(5):a016725

3. McMahon HT, Boucrot E (2011) Molecular mechanism and physiological functions of clathrin-mediated endocytosis. Nat Rev Mol Cell Biol 12(8):517–533

4. Johannes L, Parton RG, Bassereau P, Mayor S (2015) Building endocytic pits without clathrin. Nat Rev Mol Cell Biol 16(5):311–321

5. Maldonado-Baez L, Williamson C, Donaldson JG (2013) Clathrin-independent endocytosis: a cargo-centric view. Exp Cell Res 319(18):2759–2769

6. Robinson MS (2015) Forty years of Clathrin-coated vesicles. Traffic 16(12):1210–1238

7. Kirchhausen T (2009) Imaging endocytic clathrin structures in living cells. Trends Cell Biol 19(11):596–605

8. Mettlen M, Danuser G (2014) Imaging and modeling the dynamics of clathrin-mediated endocytosis. Cold Spring Harb Perspect Biol 6(12):a017038

9. Huang F, Khvorova A, Marshall W, Sorkin A (2004) Analysis of clathrin-mediated endocytosis of epidermal growth factor receptor by RNA interference. J Biol Chem 279(16):16657–16661

10. Motley A, Bright NA, Seaman MN, Robinson MS (2003) Clathrin-mediated endocytosis in AP-2-depleted cells. J Cell Biol 162(5):909–918

11. Benmerah A, Bayrou M, Cerf-Bensussan N, Dautry-Varsat A (1999) Inhibition of clathrin-coated pit assembly by an Eps15 mutant. J Cell Sci 112(Pt 9):1303–1311

12. Ford MG, Pearse BM, Higgins MK, Vallis Y, Owen DJ, Gibson A, Hopkins CR, Evans PR, McMahon HT (2001) Simultaneous binding of PtdIns(4,5)P2 and clathrin by AP180 in the nucleation of clathrin lattices on membranes. Science 291(5506):1051–1055

13. Damke H, Baba T, Warnock DE, Schmid SL (1994) Induction of mutant dynamin specifically blocks endocytic coated vesicle formation. J Cell Biol 127(4):915–934

14. Marks B, Stowell MH, Vallis Y, Mills IG, Gibson A, Hopkins CR, McMahon HT (2001) GTPase activity of dynamin and resulting conformation change are essential for endocytosis. Nature 410(6825):231–235

15. Macia E, Ehrlich M, Massol R, Boucrot E, Brunner C, Kirchhausen T (2006) Dynasore, a cell-permeable inhibitor of dynamin. Dev Cell 10(6):839–850

16. von Kleist L, Stahlschmidt W, Bulut H, Gromova K, Puchkov D, Robertson MJ, MacGregor KA, Tomlin N, Pechstein A, Chau N, Chircop M, Sakoff J, von Kries JP, Saenger W, Krausslich HG, Shupliakov O, Robinson PJ, McCluskey A, Haucke V (2011) Role of the clathrin terminal domain in regulating coated pit dynamics revealed by small molecule inhibition. Cell 146(3):471–484

17. Robinson MS, Sahlender DA, Foster SD (2010) Rapid inactivation of proteins by rapamycin-induced rerouting to mitochondria. Dev Cell 18(2):324–331

18. Boucrot E, Saffarian S, Zhang R, Kirchhausen T (2010) Roles of AP-2 in clathrin-mediated endocytosis. PLoS One 5(5):e10597

19. Settembre C, Di Malta C, Polito VA, Garcia Arencibia M, Vetrini F, Erdin S, Erdin SU, Huynh T, Medina D, Colella P, Sardiello M, Rubinsztein DC, Ballabio A (2011) TFEB links autophagy to lysosomal biogenesis. Science 332(6036):1429–1433

20. Aguet F, Upadhyayula S, Gaudin R, Chou YY, Cocucci E, He K, Chen BC, Mosaliganti K, Pasham M, Skillern W, Legant WR, Liu TL, Findlay G, Marino E, Danuser G, Megason S, Betzig E, Kirchhausen T (2016) Membrane dynamics of dividing cells imaged by lattice light-sheet microscopy. Mol Biol Cell 27(22):3418–3435

21. Boucrot E, Kirchhausen T (2007) Endosomal recycling controls plasma membrane area during mitosis. Proc Natl Acad Sci U S A 104(19):7939–7944

22. Tacheva-Grigorova SK, Santos AJ, Boucrot E, Kirchhausen T (2013) Clathrin-mediated endocytosis persists during unperturbed mitosis. Cell Rep 4(4):659–668

23. Coupin GT, Muller CD, Remy-Kristensen A, Kuhry JG (1999) Cell surface membrane homeostasis and intracellular membrane traffic balance in mouse L929 cells. J Cell Sci 112(Pt 14):2431–2440

24. Coller HA, Sang L, Roberts JM (2006) A new description of cellular quiescence. PLoS Biol 4(3):e83

25. Boucrot E, Howes MT, Kirchhausen T, Parton RG (2011) Redistribution of caveolae during mitosis. J Cell Sci 124(Pt 12):1965–1972

26. McCluskey A, Daniel JA, Hadzic G, Chau N, Clayton EL, Mariana A, Whiting A, Gorgani NN, Lloyd J, Quan A, Moshkanbaryans L, Krishnan S, Perera S, Chircop M, von Kleist L, McGeachie AB, Howes MT, Parton RG, Campbell M, Sakoff JA, Wang X, Sun JY, Robertson MJ, Deane FM, Nguyen TH, Meunier FA, Cousin MA, Robinson PJ (2013) Building a better dynasore: the dyngo compounds potently inhibit dynamin and endocytosis. Traffic 14(12):1272–1289

27. Ehrlich M, Boll W, Van Oijen A, Hariharan R, Chandran K, Nibert ML, Kirchhausen T (2004) Endocytosis by random initiation and stabilization of clathrin-coated pits. Cell 118(5):591–605

28. Rappoport JZ, Simon SM (2008) A functional GFP fusion for imaging clathrin-mediated endocytosis. Traffic 9(8):1250–1255

29. Gaidarov I, Santini F, Warren RA, Keen JH (1999) Spatial control of coated-pit dynamics in living cells. Nat Cell Biol 1(1):1–7

30. Merrifield CJ, Feldman ME, Wan L, Almers W (2002) Imaging actin and dynamin recruitment during invagination of single clathrin-coated pits. Nat Cell Biol 4(9):691–698

31. Jaqaman K, Loerke D, Mettlen M, Kuwata H, Grinstein S, Schmid SL, Danuser G (2008) Robust single-particle tracking in live-cell time-lapse sequences. Nat Methods 5(8):695–702

32. Quan A, McGeachie AB, Keating DJ, van Dam EM, Rusak J, Chau N, Malladi CS, Chen C, McCluskey A, Cousin MA, Robinson PJ (2007) Myristyl trimethyl ammonium bromide and octadecyl trimethyl ammonium bromide are surface-active small molecule dynamin inhibitors that block endocytosis mediated by dynamin I or dynamin II. Mol Pharmacol 72(6):1425–1439

33. Hill TA, Gordon CP, McGeachie AB, Venn-Brown B, Odell LR, Chau N, Quan A, Mariana A, Sakoff JA, Chircop M, Robinson PJ, McCluskey A (2009) Inhibition of dynamin mediated endocytosis by the dynoles--synthesis and functional activity of a family of indoles. J Med Chem 52(12):3762–3773

34. Robertson MJ, Hadzic G, Ambrus J, Pome DY, Hyde E, Whiting A, Mariana A, von Kleist L, Chau N, Haucke V, Robinson PJ, McCluskey A (2012) The Rhodadyns, a new class of small molecule inhibitors of Dynamin GTPase activity. ACS Med Chem Lett 3(5):352–356

35. Gibson AE, Noel RJ, Herlihy JT, Ward WF (1989) Phenylarsine oxide inhibition of endocytosis: effects on asialofetuin internalization. Am J Phys 257(2 Pt 1):C182–C184

36. Heuser JE, Anderson RG (1989) Hypertonic media inhibit receptor-mediated endocytosis by blocking clathrin-coated pit formation. J Cell Biol 108(2):389–400

37. Larkin JM, Brown MS, Goldstein JL, Anderson RG (1983) Depletion of intracellular potassium arrests coated pit formation and receptor-mediated endocytosis in fibroblasts. Cell 33(1):273–285

38. Schlegel R, Dickson RB, Willingham MC, Pastan IH (1982) Amantadine and dansylcadaverine inhibit vesicular stomatitis virus uptake and receptor-mediated endocytosis of alpha 2-macroglobulin. Proc Natl Acad Sci U S A 79(7):2291–2295

39. Wang LH, Rothberg KG, Anderson RG (1993) Mis-assembly of clathrin lattices on endosomes reveals a regulatory switch for coated pit formation. J Cell Biol 123(5):1107–1117

40. Boucrot E, Ferreira AP, Almeida-Souza L, Debard S, Vallis Y, Howard G, Bertot L, Sauvonnet N, McMahon HT (2015) Endophilin marks and controls a clathrin-independent pathway of endocytosis. Nature 517(7535):460–465

Chapter 4

Assaying the Contribution of Membrane Tension to Clathrin-Mediated Endocytosis

Steeve Boulant

Abstract

Nowadays, live fluorescent microscopes allow us to study the dynamics of cellular processes in living cells with high spatial and temporal resolution. Since the implementation of this methodology to the field of clathrin-mediated endocytosis (CME), this approach has revolutionized our molecular understanding of clathrin-driven cellular uptake. Conventional live cell microscopy approaches are used to determine the precise functions of specific proteins or lipids in orchestrating CME. Here, we will describe, in depth, the procedure to investigate the contribution of membrane tension in regulating clathrin-dependent endocytosis. We will explain two alternative methods to manipulate membrane tension while performing live fluorescence microscopy: cellular swelling through osmotic shock and cellular stretching of cells grown on stretchable silicon inserts.

Key words Clathrin-mediated endocytosis, CME, Live confocal microscopy, Osmotic shock, Cellular stretching, Membrane tension

1 Introduction

CME has been historically discovered by electron microscopy approaches. This technique has allowed us to visualize for the first time coated structures at the cellular plasma membrane [1–4]. Following the identification of clathrin as the building block for the formation of these endocytic structures [5–8], bulk internalization assays have been performed to identify cargo molecules and cellular receptors for which internalization was dependent on clathrin [9, 10]. With the implementation of fluorescence microscopy [11] to study live organisms by Stanislaus von Prowazek in 1914 and the development of green fluorescent protein (GFP) [12], our understanding of cellular processes has significantly changed. We now can gain spatiotemporal information and study when and where a dynamic molecular event takes place within a cell. The implementa-

---

The original version of this chapter was revised. A correction to this chapter can be found at https://doi.org/10.1007/978-1-4939-8719-1_19

Laura E. Swan (ed.), *Clathrin-Mediated Endocytosis: Methods and Protocols*, Methods in Molecular Biology, vol. 1847, https://doi.org/10.1007/978-1-4939-8719-1_4, © Springer Science+Business Media, LLC, part of Springer Nature 2018

tion of this technology in the field of CME has been a major break-through and instrumental to our current understanding [13–17]. It allowed us to unravel that clathrin, along with multiple adaptor and regulatory proteins assemble in a precise and concerted manner to form a clathrin-coated pit (CCP) [18–22]. Assembly of this CCP provides the energy necessary to invaginate the plasma membrane. Studies of endocytosis of both large cargo [23] and large flat clathrin array [24] and of the endocytic mechanisms in yeasts [25] have proposed that membrane tension contributes to CME. In the case of large cargo, it has been suggested that actin provides the additional energy necessary to bend the plasma membrane around these cargos [23]. In the absence of actin, endocytosis of large cargo is impaired. In the case of yeast where CME is strictly dependent on actin, it was proposed that actin provides the additional forces to compensate the natural high turgor pressure inside the yeast cells [25–27]. In these cells, the internal pressure opposes membrane deformation and increases the amount of force that has to be generated to form the endocytic structure.

Here we have adapted methods to manipulate membrane tension to directly assay in live cell its contribution in regulating CME [28]. This approach allows us to investigate the influence of membrane tension on both the dynamic of clathrin coated pit formation and their dependency on actin. First, we have used osmotic shock to induce cellular swelling and increase plasma membrane tension. Second, we have successfully developed microscopy compatible silicone inserts where cells can be grown. By stretching these silicon inserts with a micromanipulator, we can physically increase tension of the plasma membrane and address the impact of membrane tension on CME, in live conditions under the microscope.

This chapter will describe the step by step procedure to (1) manipulate membrane tension using the two above-listed methods and (2) to monitor its impact on the dynamic of CCP formation and its actin dependency.

2 Materials

All solutions are prepared with ultrapure deionized water from a Merck Millipore Milli-Q water system (or equivalent). Water resistivity must be 18 MΩ.cm at 25°C, or higher. All solutions are stored at room temperature, unless otherwise stated. Waste disposal must follow local regulations.

2.1 Cell Culture

1. African green monkey kidney cells (BSC-1) expressing the clathrin adaptor protein AP2 fused to the green fluorescent protein (AP2-GFP) (*see* **Note 1**).

2. Madin-Darby canine kidney cells (MDCK) expressing AP2-GFP (*see* **Note 1**).

3. G418 (50 mg/mL stock concentration).

4. Jasplakinolide (1 mM stock concentration in DMSO).

5. Cell culture medium for BSC-1 and MDCK cells: Dulbecco's modified Eagle's medium (DMEM) supplemented with 10% fetal bovine serum and 1% penicillin–streptomycin.

6. Live cell imaging medium: phenol red-free DMEM containing 5% fetal bovine serum. The use of antibiotic is not necessary as cells are discarded after imaging.

7. Cell culture grade trypsin 0.05%–EDTA.

8. Sterile PBS: (in 800 mL of distilled water dissolve, 8 g of NaCl, 0.2 g of KCl, 1.44 g of Na_2HPO_4, 0.24 g of KH_2PO_4. Adjust the pH to 7.4 with HCl. Add distilled water to a final volume of 1 L. Sterilize by autoclaving (20 min, 121 °C, liquid cycle). Store at room temperature.

9. Cell culture plastics (T25 cell culture flasks or cell culture dishes, sterile plastic pipettes), cell culture liquid waste vacuum pump system.

10. Temperature controlled water bath.

11. Cells are grown at 37 °C and in the presence of humidified 5% CO_2

2.2 Osmotic Swelling

1. Hypo-osmotic imaging medium: dilute live cell imaging medium: prepared by mixing 100, 80, 70, or 50% of live cell imaging medium with 0, 20, 30, or 50% water containing 5% fetal bovine serum.

2.3 Controlled Cellular Stretching

1. Polydimethylsiloxane (PDMS) membrane of 127 μm in thickness (Thickness in inches: 0.005″); Specialty Silicone Products (*see* **Note 2**).

2. Acetone.

3. Optically flat-ended fused silica rod (Techspec, Edmund Optics).

4. PDMS Sylgard 184 silicone elastomer.

5. 50 μg/mL fibronectin at prepared in water.

6. Linear translation stage that features movement capability across one axis (*see* **Note 3**).

7. Vacuum chamber.

8. Laboratory oven which can reach 65 °C.

2.4 Live-Cell Confocal Microscopy

1. 35 mm imaging dishes with glass bottom coverslips.

2. Spinning-disc fluorescent confocal microscope with a ×63 objective lens, autofocus module and an environment-controlled incubation chamber.

3. Spherical-aberration-correction device (SAC) (*see* **Note 2**).

3 Methods

3.1 Osmotic Swelling

In this first approach, cellular membrane tension will be increased by inducing cellular osmotic swelling by incubating cells in a hypo-osmotic buffer. The effect of increasing membrane tension on the dynamics of clathrin coated pits and their dependency to actin will be investigated by live confocal microscopy.

1. BSC-1 or MDCK cells stably expressing AP2-GFP are grown in a T25 cell culture flask in cell culture medium containing 400 µg/mL of G418 in cell culture incubator at 37 °C and 5% CO_2.

2. The day before the osmotic swelling experiment, confluent the T25 flask of BSC-1 or MDCK cells are removed from incubator. Culture medium is discarded with sterile plastic pipette attached to a cell culture liquid waste vacuum pump system. 5 mL of sterile PBS is added to the T25 flask. The cells are washed by gentle rocking to spread the PBS on the cell culture layer. PBS is removed with waste vacuum pump system. The PBS washing step is performed twice. After removal of the PBS, 2 mL of trypsin 0.05%–EDTA are added to the cells and the T25 flask is transferred to the cell culture incubator. When cells are detached from the cell culture flask (around 2 min) (see **Note 4**), 3 mL of fresh culture medium are added to the flask to inactivate the trypsin. Cells are dissociated by pipetting up and down with a 10 mL plastic pipette.

3. Cells are counted using a hemocytometer according to manufacturer's instructions.

4. 1×10^5 BSC1 or MDCK cells resuspended into 2 mL of cell culture medium are plated in a 35 mm imaging dish. Cells are incubated overnight in cell culture incubator.

5. The day of the osmotic swelling experiment, prewarm the microscope at 37°C for at least 1–2 h prior to imaging (see **Note 5**).

6. Hypo-osmotic imaging media is prepared fresh by diluting imaging cell culture medium with water containing 5% FBS. Hypo-osmotic medium is prepared in 15 mL conical tubes in sterile conditions. To prepare the different hypo-osmotic media follow Table 1 below. Media are then kept in a water bath at 37°C.

7. Prior to the osmotic swelling experiment, 35 mm imaging dishes are removed from the incubator. Cell culture medium is discarded and the cell are washed twice with 2 mL sterile PBS as described in point 2.

8. PBS is removed and replaced with 5 mL of prewarmed hypo-osmotic medium. Cells are place in the incubator for

Table 1
Dilution of hypo-osmotic media

Dilution factor	Imaging culture medium (mL)	Water containing 5% FBS (mL)	Final volume
10%	9	1	10
20%	8	2	10
30%	7	3	10
40%	6	4	10
50%	5	5	10

10–15 min. Only one condition is prepared at each time. After imaging the next condition is prepared.

9. Cells are removed from the incubator, the hypo-osmotic medium is removed from the 35 mm imaging dishes and replaced with 2 mL of fresh prewarmed hypo-osmotic medium.

10. The 35 mm imaging dish is placed on the sample holder of the preheated microscope and the dynamics of clathrin coated pits is monitored.

11. Cells are imaged under the microscope at a rate of 1 image every 3 s (*see* **Note 6**).

12. To monitor the actin dependency of clathrin-coated pits, the 2 mL of imaging medium or hypo-osmotic imaging medium are removed from the 35 mm imaging dish and replaced by 2 mL of fresh imaging medium or hypo-osmotic imaging medium containing 1 µM of jasplakinolide (*see* **Note 7**). Cells are incubated for 10 min in the presence of jasplakinolide and imaged with an imaging rate of 1 picture every 3 s (*see* **Note 8**).

3.2 Controlled
Cellular Stretching

1. Cut a 1.5 × 1.5 cm square of PDMS membrane from the PDMS sheet.

2. The PDMS square is washed in a 10 cm plastic dish containing 10 mL of Acetone for 5 min. The PDMS square is removed, rinsed with milliQ water and further washed in a 10 cm plastic dish containing 10 mL of 70% ethanol. Finally, the PDMS membrane is extensively washed with milliQ water and patted dry between two pieces of precision wipe paper (Kimtech/ Kimwipes paper or similar).

3. The PDMS membrane is installed at the bottom of a 35 mm plastic dish.

4. In a 15 mL conical tube, add 9 mL of PDMS Sylgard® part A and 1 mL of PDMS Sylgard® part B. Mix extensively with a

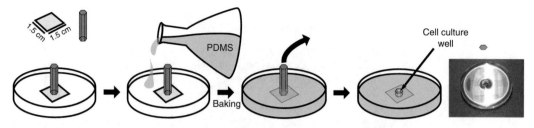

Fig. 1 Casting a PDMS cell culture chamber device. The base of the imaging device is a small sheet of PDMS membrane, while the other walls are cast from PDMS. The cell culture area is defined by the well which has been formed by removing the silica rod

2 mL plastic pipette (or equivalent) by stirring directly into the conical tube (*see* **Note 9**).

5. The 15 mL conical tube containing the mixed PDMS (*without the cap*) is placed in an adapted tube holder. The holder in placed into a vacuum chamber and vacuum is created for at least 20–30 min in order to degas the PDMS solution (*see* **Note 10**).

6. The degassed PDMS solution is then gently poured on the PDMS membrane located at the bottom the 35 mm petri dish.

7. The optically flat-ended fused silica rod is then placed vertically in the 35 mm dish and firmly pressed down against the PDMS sheet. This is used to cast the lateral walls of the PDMS chamber (*see* **Note 11**) (*see* Fig. 1).

8. The 35 mm dish containing PDMS and silica rod are then gently transferred to a laboratory oven. The PDMS is then cured for 2–4 h at 65 °C.

9. After curing, the PDMS chamber is removed from the oven and the silica rod is removed from the PDMS casting. The rod needs to be removed in a straight manner in order to not damage the bottom of the chamber (you should feel some resistance as the silicone will slightly stick to the rod). The PDMS insert is then removed from the 35 mm plastic dish either by scooping it out or by breaking the plastic dish (*see* Fig. 1).

10. The PDMS insert is then washed twice in 70% ethanol. A final wash is performed in milliQ water.

11. The PDMS insert is then coated with fibronectin to promote cell adhesion to the PDMS. For this, the well which has been formed by removing the silica rod (*see* Fig. 1) is then filled with 500 μL of 50 μg/mL fibronectin solution. The PDMS insert with the fibronectin solution is incubated overnight at 4 °C. Following overnight coating, the fibronectin solution is removed and the fibronectin-coated PDMS insert is washed with sterile PBS.

12. The day before the controlled stretching experiment, confluent a T25 flask of BSC-1 or MDCK cells is removed from

incubator. Culture medium is discarded with sterile plastic pipette attached to a cell culture liquid waste vacuum pump system. 5 mL of sterile PBS is added to the T25 flask. The cells are washed by gentle rocking to spread the PBS on the cell culture layer. PBS is removed with waste vacuum pump system. The PBS washing step is performed twice. After removal of the PBS, 2 mL of trypsin 0.05%–EDTA are added to the cells and the T25 flask is added to the cell culture incubator. When cells are detached from the cell culture flask (around 2 min) (*see* **Note 4**), 3 mL of fresh culture medium are added to the flask. Cells are dissociated by pipetting up and down with a 10 mL plastic pipette.

13. Cells are counted using a hemocytometer according to manufacturer instructions.

14. 1×10^4 BSC1 or MDCK cells resuspended into 0.5 mL of cell culture medium are plated into the well that has been cast by the silica rod. Cells are incubated overnight in the cell culture incubator.

13. The day of the stretching experiment, prewarm both the microscope at 37 °C and the stretching device for at least 1–2 h prior to imaging (*see* **Note 5**).

15. Mount the PDMS insert on the stretching device and clamp it, one side to the fixed module and the other side to the mobile module (*see* **Note 3**).

16. Remove the 0.5 mL cell culture medium from the PDMS insert, wash the cell twice with sterile PBS and replace with 37 °C prewarmed 0.5 mL of live cell imaging medium.

17. Perform microscopy to monitor the dynamics of clathrin coated pits in a relaxed (nonstretched) condition or immediately after ~25% mechanical stretching in length along one horizontal axis. In order to monitor the dynamics of clathrin-coated pit formation through the PDMS insert, set the SAC device to a value of 0. This will allow for the correction of the spherical aberration created by the PDMS sheet. If your microscope is not equipped with a SAC device, imaging through the 127 μm PDMS sheet will be extremely challenging with poor signal and very little contrast, making it difficult to follow live the dynamics of clathrin structures. In this case, I recommend to use thinner PDMS sheet or to prepare your own PDMS sheet (*see* **Note 12**).

18. Cells are imaged under the microscope with an imaging rate of 1 picture every 3 s.

19. To monitor the actin dependency of clathrin-coated pits, live cell imaging medium is removed and replaced by 0.5 mL of fresh imaging medium containing 1 μM of jasplakinolide (*see* **Note 7**). Cells are incubated for 10 min in the presence of

jasplakinolide and imaged with an imaging rate of 1 picture every 3 s (*see* **Note 13**). Cells are imaged in relaxed (non-stretched) or stretched conditions.

20. Advanced tracking and analysis of the dynamics of clathrin coated pits can be performed using the Matlab-based protocols [29, 30]. For simple tracking and analysis of clathrin structures the TrackMate plugin from the ImageJ freeware represents a good alternative (*see* **Note 14**).

4 Notes

1. Fluorescent clathrin-coated pits can be detected 24–48 h post-transfection of BSC-1 or MDCK cells with a plasmid encoding the AP2-GFP construct [31]. Following transient transfection, a significant fraction of the AP2-GFP expressing cells (up to 50%) will display cytosolic staining and will not have AP2-GFP incorporated into clathrin-coated structures (or very little). Depending of the cell lines used to monitor the dynamics of clathrin-coated structures, this fraction can be more or less abundant. We recommend establishing a stably expressing AP2-GFP cell line through geniticin selection to obtain a population where most of the cells have the clathrin adaptor protein AP2 fused to GFP incorporated into clathrin-coated structures. Ultimately, performing cell sorting and isolating cells with low AP2-GFP expression level will allow the isolation of a cellular clone with very little cytosolic AP2-GFP background. This will significantly help for tracking individual clathrin structures.

2. The PDMS membrane will constitute the bottom of the stretchable insert. Live cell monitoring of clathrin-coated pit dynamics will be performed by imaging through this PDMS membrane. It is critical that the thinnest possible PDMS is used to improve the quality of imaging. Indeed, when imaging through the PDMS membrane severe spherical aberration (SA) will be created. SA will cause the image to be blurry with very little contrast. SA is caused when imaging is performed through material which has a different refractive index than the lens manufacturer's specification. SA becomes worse the deeper we are looking into a sample. In the context of the PDMS stretchable insert, the PDMS membrane has a different refractive index compared to glass. Importantly, given the thickness of the PDMS membrane, the SA is even more a problem. In order to partially remove SA, our microscope is equipped with a spherical-aberration-correction (SAC) device. This unit contains a series of motorized lenses that can be moved to correct the SA and generate a better focus image. If your microscope

is not equipped with a SAC, try to use PDMS membrane as thin as possible. In **Note 12** an alternative approach is described to make your own PDMS membrane with a thickness close to 20–30 μm.

3. The stretching device is a homemade linear Translation Stretching stage (from Edmund Optics). In brief, a fix plate is attached from the bottom to an on-axis translation stage. The stretchable insert is mounted on top of the device overlapping the fix plate and the translational stage. The insert is held in place with tie down table clamps (*see* Fig. 2). Commercial devices allowing controlled stretching are now becoming available. (For example: CellScale Biomaterial testing or STREX Inc.) However, the PDMS membrane used to make the stretchable insert by these companies is usually thick (around 250 μm). This should be taken in consideration for imaging. Home-made stretchable inserts compatible with commercially available stretching devices might be made following the guidelines provided in **Note 12**.

4. Trypsinization of BSC-1 cells takes around 1–2 min at 37°C. In the case of MDCK cells, if they have been growing for a couple of days and if the cells became very confluent, MDCK cells will polarize and develop tight junctions between each other. This polarization will make MDCK cells much more difficult to dissociate with trypsin. In this case, cell dissociation with trypsin-EDTA at 37°C can take up to 10 min.

5. Prewarming the microscope system at least 1–2 h prior to imaging will prevent thermal drift during the imaging of clathrin dynamics. When the temperature of a microscope is moved from room temperature to 37°C, the metal frame and optic parts dilate. This dilatation results in fine movements between the optics and samples resulting in a loss of focus while imaging.

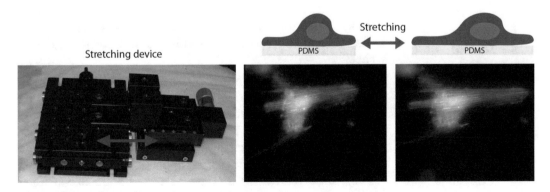

Fig. 2 Cell stretching device and typical images obtained from stretched cells

6. With increasing dilution factor of the imaging medium (more hypo-osmotic medium), the average lifetime of clathrin-coated pit should increase. The 10–15 min preincubation time in the hypo-osmotic medium could be performed directly in the microscope incubation chamber. This preincubation step is necessary for hypo-osmotic shock to induce cell swelling, to increase membrane tension, and to affect CME. The 10–15 min preincubation time has been determined to be optimum for BSC-1 cells. If using a different cell line, the optimum time post hypo-osmotic shock when the greatest impact on CME is observed should be experimentally determined.

7. Replacing the imaging medium or hypo-osmotic imaging medium with medium containing 1 μM jasplakinolide can be done directly under the microscope. Make sure to use pre-warmed medium at 37°C to avoid thermal drift. We prefer to use jasplakinolide instead of latrunculin since jasplakinolide induces/stabilizes actin filaments. Latrunculin depolymerizes F-actin and quickly induces cell rounding. Prolonged treatment with jasplakinolide (30 min to 1 h) also induces cell rounding but to a lesser extent compared to latrunculin.

8. With increasing dilution factor of the imaging medium (more hypo-osmotic medium) and in the presence of 1 μM jasplakinolide, the number of clathrin coated pits arrested at the plasma membrane (no longer dynamic) should increase.

9. Perfect mixing of the PDMS base part A and the curing agent part B is key to ensuring uniform polymerization of the silicone.

10. During degassing, fine air bubbles should form in the PDMS and move to the surface of the PDMS solution. When no more air bubbles are detectable, correct degassing has been achieved. If no air bubbles are detected, it is likely the vacuum is not sufficient. Try increasing the vacuum pressure.

11. At this stage the PDMS solution should be degassed. If you notice small bubbles after pouring, the 35 mm dish containing the liquid PDMS and the silica rod can be placed for an additional 10–20 min in the vacuum chamber. Make sure that the silica rod remains vertical and firmly pressed against the PDMS sheet.

12. If your microscope is not equipped with a SAC device, you might not be able to look through the 127 μm thick PDMS membrane. It will be necessary to use prepare thinner PDMS membranes which can be easily done in the lab with classical soft-lithography equipment and a plasma cleaner for electron microscopy (EM facilities are equipped with plasma cleaners). The PDMS membrane is prepared on a 2.5 cm silicon wafer which represents an extremely clean and flat surface. The wafer

is first cleaned with freshly prepared piranha solution (3:1 H_2SO_4–H_2O_2 (30%)) for 1 h at room temperature under the chemical hood. The piranha solution is extremely dangerous and needs be handled with great care. The wafer is washed extensively with deionized water and then dried from any moisture by heating the wafer in an oven at 120°C for 15 min. Finally, the wafer is treated for 5 min in the plasma cleaner. To better remove the PDMS from the wafer a sacrificial layer will be added on top of the wafer by spreading AZ4562 photoresist on the surface of the wafer by spin coating. Spin coating is a technique that uses a centrifuge-like unit that allows for very fast dispersion of a liquid onto a surface to create very thin uniform layers. The wafer is mounted in the center of the spin coater rotor and 1 mL of AZ4562 photoresist is added on top the wafer (for a wafer of 2.5 cm diameter). Spin coating takes place in two steps. First spin: 500 rpm for 10 s. Second spin: 5000 rpm for 30 s. Both with an acceleration of 300 rpm/s. The coated wafer is then baked for 2 min at 100°C. The PDMS membrane can now be spin coated on the wafer having the sacrificial layer. PDMS Sylgard® 184 is mixed and degassed as described above and deposited in the center of wafer located in the spin coater. By adjusting the spinning speed and time, it is possible to control the thickness of the PDMS layer that will be performed during the spin coating process. As indicated, 5 min spinning at 5000 rpm will provide a PDMS layer of around 5 μm thick. Many tables correlating spinning time and speed with the final thickness obtained of the PDMS membrane can be found online. After baking the PDMS (as described above), the sacrificial layer is removed by immerging the PDMS coated wafer in acetone. After 2–5 min, the AZ4562 photoresist will dissolve and release the PDMS membrane which will float on top of the acetone. This PDMS membrane can now be collected gently and used to pour the stretchable PDMS insert (*see* above).

13. With increasing linear stretching (more membrane tension), the average lifetime of clathrin-coated pits should slightly increase. Importantly, in the presence of 1 μM jasplakinolide under stretched conditions the number of clathrin coated pits arrested at the plasma membrane (no longer dynamic) should be significantly higher.

14. Tracking of clathrin structures with ImageJ (TrackMate)

The ImageJ plugin TrackMate can be used as an easy tool to detect and segment spherical objects like clathrin structures and track them over time. In most cases the preprocessing of the images is not necessary although a background correction (e.g., rolling-ball background correction) can help limit false positive object detection. After importing your movie into

ImageJ (usually a .Tiff file), start the TrackMate plugin. First check the calibration settings (pixel width and height, voxel depth, and time interval) which are taken from the metadata of your images. The selection of the detector for detection and segmentation of the objects depends on the size of your structures in pixels. In the case of clathrin coated pits, in our setup, we use the DoG detector which is well adapted for small objects (pixel diameter below 5). If your cells of interest have a lot of large clathrin structures (plaques) or depending of your imaging/camera setup, you might have to use the LoG detector, better adapted for larger objects (pixel diameter between 5 and 20). In any case, the settings of the detector and the thresholding need to be adjusted for best detection. Check if all clathrin structures are detected by looking at the preview. Select the HyperStack Displayer as a viewer. In the following section, you can choose between different filters to exclude some spots based on their position in x, y, time, intensity, contrast, or estimated diameter. If you do not want to use any filter, you can continue by selecting a tracker. In the case of clathrin structures, there are very little splitting or merging events, as such we use the simple LAP (Linear Assignment Problem) tracker. You will then have to set the linking maximal distance, the gap-closing maximal distance and the gap-closing maximal frame gap. Since the clathrin structures do not move much in x and y position, you need to choose a small linking distance in the range of the spot size. We set the gap-closing maximal frame gap to 0, assuming that there are no gaps in the clathrin tracks. After the tracking you should check the tracking accuracy and adjust the parameters if necessary. You can then export the tracking information as a table with the analysis button. This will give you three tables with all the tracking information: statistics over the spots in tracks, the links in tracks and the tracks. There is the option to directly create plots using different features in ImageJ as well as to capture an overlay of the movie and the tracks for visualization.

References

1. Friend DS, Farquhar MG (1967) Functions of coated vesicles during protein absorption in the rat vas deferens. J Cell Biol 35:357–376 The Rockefeller University Press

2. Heuser JE, Reese TS (1973) Evidence for recycling of synaptic vesicle membrane during transmitter release at the frog neuromuscular junction. J Cell Biol 57:315–344 The Rockefeller University Press

3. Maupin P, Pollard TD (1983) Improved preservation and staining of HeLa cell actin filaments, clathrin-coated membranes, and other cytoplasmic structures by tannic acid-glutaraldehyde-saponin fixation. J Cell Biol 96:51–62 The Rockefeller University Press

4. ROTH TF, PORTER KR (1964) Yolk protein uptake in the oocyte of the mosquito aedes aegypti. L. J Cell Biol 20:313–332 The Rockefeller University Press

5. Heuser J (1980) Three-dimensional visualization of coated vesicle formation in fibroblasts. J Cell Biol 84:560–583 The Rockefeller University Press

6. Kirchhausen T, Harrison SC, Chow EP, Mattaliano RJ, Ramachandran KL, Smart J, Brosius J (1987) Clathrin heavy chain: molecular cloning and complete primary structure. Proc Natl Acad Sci U S A 84:8805–8809 National Academy of Sciences

7. Pearse BM (1976) Clathrin: a unique protein associated with intracellular transfer of membrane by coated vesicles. Proc Natl Acad Sci U S A 73:1255–1259 National Academy of Sciences

8. Vigers GP, Crowther RA, Pearse BM (1986) Three-dimensional structure of clathrin cages in ice. EMBO J 5:529–534 European Molecular Biology Organization

9. Brown MS, Goldstein JL (1979) Receptor-mediated endocytosis: insights from the lipoprotein receptor system. Proc Natl Acad Sci U S A 76:3330–3337 National Academy of Sciences

10. Gorden P, Carpentier JL, Cohen S, Orci L (1978) Epidermal growth factor: morphological demonstration of binding, internalization, and lysosomal association in human fibroblasts. Proc Natl Acad Sci U S A 75:5025–5029 National Academy of Sciences

11. Masters BR (2001) History of the electron microscope in cell biology. Wiley, Chichester

12. Shimomura O, Johnson FH, Saiga Y (1962) Extraction, purification and properties of aequorin, a bioluminescent protein from the luminous hydromedusan, Aequorea. J Cell Comp Physiol 59:223–239

13. Ehrlich M, Boll W, Van Oijen A, Hariharan R, Chandran K, Nibert ML, Kirchhausen T (2004) Endocytosis by random initiation and stabilization of clathrin-coated pits. Cell 118:591–605

14. Gaidarov I, Santini F, Warren RA, Keen JH (1999) Spatial control of coated-pit dynamics in living cells. Nat Cell Biol 1:1–7

15. Kirchhausen T (2009) Imaging endocytic clathrin structures in living cells. Trends Cell Biol 19:596–605

16. Merrifield CJ, Feldman ME, Wan L, Almers W (2002) Imaging actin and dynamin recruitment during invagination of single clathrin-coated pits. Nat Cell Biol 4:691–698 Nature Publishing Group

17. Merrifield CJ, Perrais D, Zenisek D (2005) Coupling between clathrin-coated-pit invagination, cortactin recruitment, and membrane scission observed in live cells. Cell 121:593–606

18. Faini M, Beck R, Wieland FT, Briggs JAG (2013) Vesicle coats: structure, function, and general principles of assembly. Trends Cell Biol 23:279–288

19. Ferguson SM, De Camilli P (2012) Dynamin, a membrane-remodelling GTPase. Nat Rev Mol Cell Biol 13:75–88

20. McMahon HT, Boucrot E (2011) Molecular mechanism and physiological functions of clathrin-mediated endocytosis. Nat Rev Mol Cell Biol 12:517–533 Nature Publishing Group

21. Merrifield CJ, Kaksonen M (2014) Endocytic accessory factors and regulation of clathrin-mediated endocytosis. Cold Spring Harb Perspect Biol 6:a016733–a016733 Cold Spring Harbor Lab

22. Taylor MJ, Perrais D, Merrifield CJ (2011) A high precision survey of the molecular dynamics of mammalian clathrin-mediated endocytosis. PLoS Biol 9:e1000604 (S. L. Schmid, Ed.)

23. Cureton DK, Massol RH, Whelan SPJ, Kirchhausen T (2010) The length of vesicular stomatitis virus particles dictates a need for actin assembly during clathrin-dependent endocytosis. PLoS Pathog 6:e1001127 (J. A. T. Young, Ed.)

24. Saffarian S, Cocucci E, Kirchhausen T (2009) Distinct dynamics of endocytic clathrin-coated pits and coated plaques. PLoS Biol 7:e1000191–e1000118 (F. Hughson, Ed.)

25. Aghamohammadzadeh S, Ayscough KR (2009) Differential requirements for actin during yeast and mammalian endocytosis. Nat Cell Biol 11:1039–1042 Nature Publishing Group

26. Basu R, Munteanu EL, Chang F (2014) Role of turgor pressure in endocytosis in fission yeast. Mol Biol Cell 25:679–687 American Society for Cell Biology

27. Dmitrieff S, Nédélec F (2015) Membrane mechanics of endocytosis in cells with turgor. PLoS Comput Biol 11:e1004538 (H. Ewers, Ed.). Public Library of Science

28. Boulant S, Kural C, Zeeh J-C, Ubelmann F, Kirchhausen T (2011) Actin dynamics counteract membrane tension during clathrin-mediated endocytosis. Nat Cell Biol 13:1124–1131 Nature Publishing Group

29. Aguet F, Antonescu CN, Mettlen M, Schmid SL, Danuser G (2013) Advances in analysis of low signal-to-noise images link dynamin and AP2 to the functions of an endocytic checkpoint. Dev Cell 26:279–291

30. Cocucci E, Aguet F, Boulant S, Kirchhausen T (2012) The first five seconds in the life of a clathrin-coated pit. Cell 150:495–507

31. Boulant S, Stanifer M, Kural C, Cureton DK, Massol R, Nibert ML, Kirchhausen T (2013) Similar uptake but different trafficking and escape routes of reovirus virions and infectious subvirion particles imaged in polarized Madin-Darby canine kidney cells. Mol Biol Cell 24:1196–1207 American Society for Cell Biology

Chapter 5

Identifying Small-Molecule Inhibitors of the Clathrin Terminal Domain

Volker Haucke and Michael Krauß

Abstract

Clathrin-mediated endocytosis (CME) is a universal and evolutionarily conserved process that enables the internalization of numerous cargo proteins, including receptors for nutrients and signaling molecules, as well as synaptic vesicle reformation. Multiple genetic and chemical approaches have been developed to interfere with this process. However, many of these tools do not selectively block CME, for example by targeting components shared with clathrin-independent endocytosis pathways or by interfering with other cellular processes that indirectly affect CME.

Clathrin, via interactions of endocytic proteins with its terminal domain (TD), serves as a central interaction hub for coat assembly in CME. Here, we describe an ELISA-based, high-throughput screening method used to identify small molecules that inhibit these interactions. In addition, we provide protocols for the purification of recombinant protein domains used for screening, e.g., the clathrin TD and the amphiphysin B/C domain. The screen has been applied successfully in the past, and ultimately led to the discovery of the Pitstop® family of inhibitors, but remains in use to evaluate the inhibitory potency of derivatives of these compounds, and to screen for completely novel inhibitor families.

Key words Endocytosis, ELISA, Small-molecule inhibitor, Clathrin, Clathrin terminal domain, Clathrin box

1 Introduction

CME is a universal and evolutionary conserved process that promotes uptake of numerous cargo proteins, including receptors for nutrients such as transferrin, signaling receptors, ion channels, transporters for small metabolites, and is involved in synaptic vesicle reformation at synapses. CME has therefore been implicated in a large variety of essential cellular functions, such as the maintenance of ion and nutrient homeostasis, the modulation of signaling in neuronal and nonneuronal cells, or cell differentiation and migration during the development or organs and tissues.

The original version of this chapter was revised. A correction to this chapter can be found at
https://doi.org/10.1007/978-1-4939-8719-1_19

Laura E. Swan (ed.), *Clathrin-Mediated Endocytosis: Methods and Protocols*, Methods in Molecular Biology, vol. 1847,
https://doi.org/10.1007/978-1-4939-8719-1_5, © Springer Science+Business Media, LLC, part of Springer Nature 2018

To dissect the impact of CME on such cellular functions researchers have developed a variety of genetic and chemical tools to manipulate this process. One of the first genetic approaches made use of a mutation of dynamin2 that abolishes its GTPase activity [1]. Dynamin is required for vesicle fission at late stages in CME. Overexpression of mutant dynamin in cells stalls CME at the stage of Ω-shaped invaginated pits coated with clathrin and endocytic proteins such as the heterotetrameric AP-2 adaptor. As dynamin is not only involved in CME but also supports clathrin-independent modes of endocytosis alternative dominant-negative approaches were developed to selectively inhibit CME. Many proteins involved in CME harbor multiple binding sites for clathrin and/or AP-2, i.e., the so-called B/C domain of amphiphysin [2, 3] or the carboxyl-terminal region of AP180 [4]. Overexpression of these largely soluble domains in cells sequesters clathrin and AP-2 in the cytoplasm and thus induces a block in CME. A widely used alternative to the overexpression of dominant negatively acting endocytic protein domains, which may adversely affect cell physiology, is the use of RNAi to deplete key endocytic proteins such as clathrin heavy chain or the μ-subunit of AP-2 from cells. As expected RNAi-mediated depletion of these proteins indeed abrogates CME [5, 6]. A key problem with any of these approaches is the long time scale (days) at which interference with CME occurs and, thus, the possibility for indirect cellular effects, e.g., caused by the redistribution of CME cargoes, the role of clathrin in mitosis, or possible compensatory upregulation of other factors. Moreover, such strategies do not allow the analysis of pathways indirectly linked to CME such as endocytic recycling of internalized cargo as CME inhibition by RNAi-mediated protein depletion or dominant-negative interference is essentially irreversible.

To enable the acute reversible perturbation of CME chemical inhibitors have been developed. One of the first screens led to the discovery of dynasore, a noncompetitive inhibitor of dynamin1 and dynamin2 [7]. Further optimization of this compound yielded dynoles and dingos that display higher potency and less cytotoxicity [8, 9]. Among the advantages of these compounds is their rapid mode of action, and the reversibility of the treatment, while their specificity has been lately questioned [10]. Moreover, inhibition of dynamin not only blocks CME but also impacts other cellular processes as dynamin regulates other endocytosis pathways and may have secondary roles in the regulation of actin dynamics, mitosis, exocytic fusion pore dilation or closure, or cellular cholesterol homeostasis [11].

These caveats have prompted us to search for more selective inhibitors of CME. Clathrin via interactions of the N-terminal domain (TD) of its heavy chain with other endocytic proteins represents a central interaction hub within the endocytic system. The clathrin TD adopts a seven-bladed β-propeller fold that—in different grooves—provides binding sites for simple degenerate peptide motifs, including so-called clathrin-box and W-box motifs, as well as variants derived thereof [12–15]. These binding sites can function independently, and thus may allow for the cooperative association of the clathrin TD with multiple binding partners. Alternatively, single ligands of the TD may occupy more than one site, as has been observed in the case of amphiphysin [3]. We therefore designed our screen based on the hypothesis that small-molecule inhibitors that selectively perturb interactions between the clathrin TD and its ligands would impair the functionality of the clathrin coat and thereby abrogate CME. We used the amphiphysin B/C domain as a ligand for our screen as this domain displays high affinity binding to the clathrin TD via at least two binding sites.

Here, we describe the setup of our screen in detail [16], and provide protocols for the purification of the amphiphysin B/C domain and the clathrin TD used for screening.

2 Materials

Prepare all solutions in ultrapure water (18 MΩ at 25 °C). Use analytical grade reagents and prepare all solutions at 4 °C, if not indicated otherwise.

2.1 Expression of Fusion Proteins

1. 2× YT medium: Weigh 16 g of tryptone, 10 g of yeast extract, and 5 g NaCl and dissolve in 1 L of water in a 2 L flask. Mix and adjust pH to 7.0 with 5 N NaOH (*see* **Note 1**) Seal flask with aluminum foil and autoclave immediately. Store at room temperature (RT).

2. Antibiotics: Dissolve 5 g of ampicillin in 50 mL of water. Dissolve 2.5 g kanamycin in 50 mL of water. Sterilize both solutions by passing them through a 0.2 μm polyethersulfone filter using a sterile 50 mL syringe. Store solutions in 1–2 mL aliquots at −20 °C.

3. 1 M Isopropyl b-D-1-thiogalactopyranoside (IPTG). Dissolve 4.76 g of IPTG in 20 mL of water. Sterilize solution by it through a 0.2 μm polyethersulfone filter using a 50 mL syringe. Store in 1 mL aliquots at −20 °C.

4. *Escherichia coli* BL21-Codon Plus (DE3)-RP.

5. 1 M PMSF.

6. Benzonase.

7. Sonicator.

8. GST-Sepharose 4b.

9. Ni Sepharose 6 Fast Flow medium (GE Healthcare).

10. Chromatography columns.

11. 10 K sample concentrator.

12. GE Superdex 200 16/60 column.

**2.2 Purification
of His₆-Tagged
Amphiphysin B/C
Domain**

1. Resuspension buffer: 20 mM Tris pH 7.5, 150 mM NaCl, 1% CHAPS.

2. 5 M NaCl.

3. Loading buffer: 20 mM Tris pH 7.5, 500 mM NaCl.

4. Elution buffer: 20 mM Tris pH 7.5, 500 mM NaCl, 500 mM imidazole.

5. Storage buffer: 10 mM HEPES pH 7.4, 50 mM NaCl, 10% glycerol, 3 mM EDTA.

6. His_6-tagged rat amphiphysin 1 B/C domain (amino acids 250–578) in pET28a.

**2.3 Purification
of GST-Tagged Clathrin
TD**

1. Resuspension buffer: 180 mM PBS (137 mM NaCl, 2.7 mM KCl, 10 mM Na_2HPO_4, 1.8 mM KH_2PO_4 pH 7.4), 1% Triton X-100, 1 mM DTT.

2. Loading buffer: PBS.

3. Elution buffer: PBS, 20 mM reduced glutathione.

4. Storage buffer: 10 mM HEPES pH 7.4, 50 mM NaCl, 10% glycerol.

5. Human clathrin heavy chain TD (amino acids 1–364) in pGEX4T-1.

**2.4 Elisa-Based
Compound Screen**

1. Binding buffer: 10 mM HEPES pH 7.4, 50 mM NaCl, 1 mM DTT.

2. Wash buffer: 10 mM HEPES pH 7.4, 50 mM NaCl, 1 mM DTT, 0.05% Tween20.

3. Blocking buffer: 2.5% skimmed milk and 2% bovine serum albumin in binding buffer.

4. 384-well maxisorp plate (Corning: standard plate format, flat bottom, High Bind-treated surface, nonsterile).

5. Microplate washer.

6. Plate reader which reads absorbance at 450 nm.

7. Peroxidase-conjugated, rabbit polyclonal anti-GST antibody (SIGMA-ALDRICH).

8. Detection reagent: Pierce 1-Step Ultra TMB.

9. Stop solution: 2 N H_2SO_4.

3 Methods

**3.1 Expression
of Fusion Proteins**

1. Inoculate 50 mL cultures of *Escherichia coli* BL21-Codon Plus (DE3)-RP cells (*see* **Note 2**) transformed with plasmids for expression of GST-tagged human clathrin heavy chain TD (amino acids 1–364) (pGEX4T-1, ampicillin resistance) or of

His$_6$-tagged rat amphiphysin 1 B/C domain (amino acids 250–578) (pET28a, kanamycin resistance) in 2× YT medium. Add appropriate antibiotics in a 1:1000 dilution. Inoculate at 37 °C overnight.

2. Dilute overnight culture 1:100 into fresh 2× YT medium containing antibiotics (4 L for expression of GST-clathrin heavy chain TD, 2 L for expression of His$_6$-amphiphysin 1 B/C domain).

3. Inoculate bacterial cultures in flasks in an incubator at 30 °C until bacterial density has reached an OD of 0.6–0.7.

4. Cool flasks to 16 °C.

5. Add IPTG to a final concentration of 1 mM (1:1000 dilution).

6. Shake flasks in an incubator at 16 °C overnight (*see* **Note 3**).

7. Pellet cells by centrifugation at 4 °C in 500 mL buckets for 15 min at $4500 \times g$.

8. Decant supernatant and store pellets at −20 °C (*see* **Note 4**).

3.2 Purification of His$_6$-Tagged Amphiphysin B/C Domain

1. Thaw pellets and resuspend each in 20 mL of ice-cold resuspension buffer.

2. Combine resuspended pellets (40 mL) and add 1 tablet of cOmplete protease inhibitor (*see* **Note 5**), 1 mM PMSF and 200 units of benzonase (*see* **Note 6**).

3. Transfer suspensions into 50 mL Falcon tube, keep on ice and sonicate for 60 s at 50 duty and 30% power in a Branson 250D sonicator (or similar) to disrupt cells (*see* **Note 7**).

4. Add additional 350 mM NaCl (from a 5 M stock solution) and incubate on ice for 15 min.

5. Repeat sonication **step 3**.

6. Pellet cellular debris by centrifugation at 4 °C for 20 min at $50,000 \times g$.

7. Recover supernatant and apply to 9 mL of a Ni Sepharose 6 Fast Flow medium packed into a chromatography column, and preequilibrated in loading buffer (flow rate 1 mL/min) (*see* **Note 8**).

8. Wash column with 10 column volumes (90 mL; about 90 min) of loading buffer.

9. Elute in one step by passing 80 mL of 60% elution buffer (corresponding to 300 mM imidazole) through the column.

10. Collect eluate in 6 mL fractions. The protein, as monitored by absorbance at 280 nm, elutes within three successive fractions.

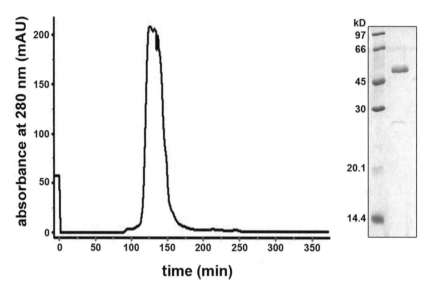

Fig. 1 Purification of the His$_6$-amphiphysin B/C domain. (left) Elution profile of the GE Superdex 200 16/60 column chromatography. (right) An aliquot of the purified His$_6$-amphiphysin B/C domain was separated by SDS-PAGE. Proteins were visualized by Coomassie Blue staining (*see* **Note 23**)

11. Pool eluates (18 mL) and concentrate sample to 2.5 mL by centrifuging in a 10 K sample concentrator at 4 °C and 4000 × *g* This step will take 30–45 min.

12. Pass concentrated sample through a 0.2 μm filter, using a 5 mL syringe (*see* **Note 9**).

13. Apply concentrated and filtered sample to a GE Superdex 200 16/60 column preequilibrated in storage buffer (flow rate 0.75 mL/min).

14. Collect eluate in 6 mL fractions. The recombinant protein, as monitored by absorbance at 280 nm, elutes in three to four successive fractions (*see* **Note 10**) (Fig. 1).

15. Determine protein concentration by measuring the absorbance at 280 nm (extinction coefficient ε_{280nm} = 11,380). The purified His$_6$-tagged amphiphysin B/C domain will have a concentration of about 2 mg/mL.

16. Freeze protein solution in 0.5 mL aliquots in liquid nitrogen and store at −80 °C.

3.3 Purification of GST-Tagged Clathrin TD

1. Thaw pellets and resuspend each in 20 mL of ice-cold resuspension buffer.

2. Combine resuspended pellets (160 mL) and add 1 tablet of cOmplete protease inhibitor (*see* **Note 4**), 1 mM PMSF and 200 units of benzonase (*see* **Note 6**).

3. Transfer suspensions into 250 mL glass beaker, place on ice and sonicate for 120 s at 50 duty and 30% power in a Branson 250D sonicator (or similar) to disrupt cells (*see* **Note 7**).

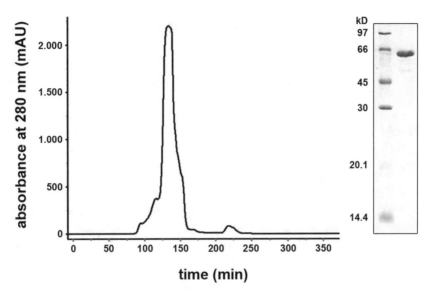

Fig. 2 Purification of the GST-clathrin TD. (left) Elution profile of the GE Superdex 200 16/60 column chromatography. (right) An aliquot of the purified GST-clathrin TD was separated by SDS-PAGE. Proteins were visualized by Coomassie Blue staining (*see* **Note 24**)

4. Pellet bacterial debris by centrifugation at 4 °C for 20 min at 50,000 × *g*.

5. Recover supernatant and apply to 8 mL of GST-Sepharose 4b material (GE Healthcare) packed into a chromatography column, preequilibrated in loading buffer. Set the flow rate to 2 mL/min (*see* **Note 11**).

6. Wash column with 10 column volumes (80 mL; about 40 min) of loading buffer.

7. Elute in one step by passing 25 mL of 100% elution buffer containing reduced glutathione through the column.

8. Collect eluate in 5 mL fractions. The protein, as monitored by absorbance at 280 nm, elutes within five successive fractions.

9. Pool eluates (25 mL) and concentrate sample to 2.5 mL by centrifuging in a 10 K sample concentrator at 4 °C and 4000 × *g*. This step will take 45–60 min.

10. Pass concentrated sample through a 0.2 μm filter, using a 5 mL syringe (*see* **Note 9**).

11. Apply concentrated and filtered sample to a GE Superdex 200 16/60 column preequilibrated in storage buffer (flow rate 0.75 mL/min).

12. Collect eluate in 6 mL fractions. The recombinant protein, as monitored by absorbance at 280 nm, elutes in three to four successive fractions (*see* **Note 10**) (Fig. 2).

13. Determine protein concentration by measuring the absorbance at 280 nm (extinction coefficient ε_{280nm} = 74,760). The purified

GST-tagged clathrin TD will have a concentration of about 6–7 mg/mL.

14. Supplement protein solution with 2 mM EDTA and 1 mM DTT, freeze in 0.5 mL aliquots in liquid nitrogen and store at −80 °C.

3.4 ELISA-Based Compound Screen to Identify Candidate Hits

First, the number of individual samples to be analyzed should be calculated: Each compound should be tested in duplicate or better triplicate (Fig. 3). In addition, positive controls (no inhibitor added: wells A1–A9 in Fig. 3) and negative controls (addition of Pitstop2: wells A4–I3 in Fig. 3; no binding ligand added: wells K1–K3 in Fig. 3) should be included to evaluate the maximal effect to be observed in the screen. If several plates are prepared, each plate should contain positive and negative controls to calculate the Z′ factor (*see* below).

1. Dilute purified His$_6$-tagged amphiphysin B/C domain to a concentration of 1.6 mg/mL in binding buffer.

2. Apply 50 μL of the diluted His$_6$-tagged amphiphysin B/C domain (corresponding to 80 μg) per well of a 384-well dish using a dispenser (wells A1–I9 and K1–K3 in Fig. 3) (*see* **Note 12**).

3. Dilute a fresh aliquot of GST-tagged clathrin TD to a concentration of 10 μg/mL in binding buffer. Apply 50 μL to three wells of the plate (wells L1–L3 in Fig. 3).

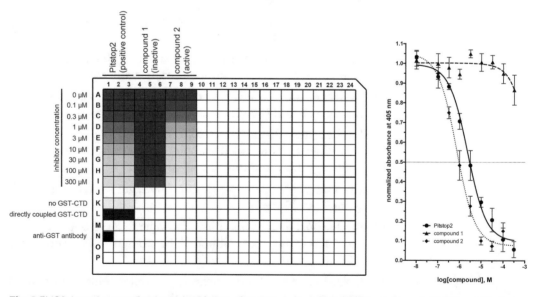

Fig. 3 ELISA-based screen for the identification of inhibitors of clathrin TD interactions. (left) Schematic representation of the organization of a 384-well microtiter plate set up to determine of dose–response curves of individual compounds. CTD, clathrin terminal domain. (right) Exemplary plots for individual compounds tested at different concentrations in the screen

4. Incubate for 1 h at room temperature (*see* **Note 13**).

5. Add 50 μL of blocking buffer per well.

6. Incubate at 4 °C overnight.

7. Remove solution using a plate washer.

8. Wash five times briefly with 100 μL of wash buffer.

9. Add 30 μL of binding buffer to each well containing coupled His$_6$-tagged amphiphysin B/C domain.

10. Add 0.4 μL of DMSO or of a 10 mM stock solution of each compound to be tested (dissolved in DMSO) (*see* **Notes 14** and **15**).

11. Add 0.4 μL of a 10 mM stock solution of Pitstop2 dissolved in DMSO to two or three wells (*see* **Note 15**).

12. Dilute a fresh aliquot of GST-tagged clathrin TD to a concentration of 80 μg/mL in binding buffer.

13. Add 10 μL of the diluted GST-clathrin TD (corresponding to 800 ng) to each well coated with His$_6$-tagged amphiphysin B/C domain except for the wells that serve as negative controls to which no binding ligand is added (*see* **Note 16**).

14. Incubate for 1 h at room temperature.

15. Wash the wells five times briefly with 100 μL of wash buffer using a microplate washer.

16. Dilute the peroxidase-coupled anti-GST antibody 1:5000 in wash buffer, thoroughly mix, and then add 50 μL to each well (*see* **Note 17**).

17. Incubate at room temperature for 15 min.

18. Wash five times in 100 μL of wash buffer as indicated above.

19. To be able to assess that the detection reaction itself works successfully add 1 μL of the diluted, peroxidase-coupled anti-GST antibody to one completely untreated well (well N1 in Fig. 3) (*see* **Note 18**).

20. Add 50 μL of detection reagent to each well and incubate the plate at room temperature for 15 min.

21. Stop detection reaction by adding 50 μL of the stop solution (*see* **Note 19**).

22. Determine absorbance at 450 nm for each well using an appropriate plate reader.

23. Determine mean absorbance for all wells treated equally in duplicate or triplicate.

24. Determine Z′ factor from positive and negative controls on each plate (*see* **Note 20**).

25. Subtract background (measured from samples not supplemented with GST-tagged clathrin-TD; *see* **step 12** and **Note 16**) from all averaged absorbances.

26. Normalize all mean absorbance values to the value determined for the positive control treated with DMSO only (Fig. 3).

3.5 Dose–Response Curves for Candidate Hits

Hits identified in the primary screen should be further validated, before testing them in more elaborate cellular assays. Therefore, in a secondary screen different concentrations of the identified compounds are applied to evaluate their potency to disrupt clathrin TD interactions. This secondary screen follows essentially the same protocol as described above (*see* Subheading 3.4) but includes two important modifications.

1. Serial dilution of candidate compounds:
   Prepare a 30 mM stock solution of the individual compound in DMSO, and then dilute this solution to final concentrations of 10 mM, 3 mM, 1 mM, 0.3 mM, 0.1 mM, 0.03 mM, and 0.01 mM. Use these solutions to apply different concentrations of inhibitor during the binding reaction (**step 9** in Subheading 3.4) (*see* **Note 21**).

2. To allow for the detection of smaller effects, as one would expect for lower inhibitor concentrations, the peroxidase-coupled anti-GST antibody is allowed to bind for 40 min instead of 15 min (**step 16** in Subheading 3.4).

The screen described above will allow for the identification of the most potent inhibitors in vitro. However, subsequent rounds of testing are necessary to evaluate if the identified compounds are also active in living cells, and which mechanism may underlie their action. First, the compounds' capability to block the uptake of radiolabeled or fluorescently labeled transferrin into cells should be analyzed (*see* **Note 22**). Second, effects on coated pit dynamics should be investigated using total internal reflection (TIRF) microscopy on living fibroblasts stably expressing a fluorescent protein-coupled clathrin or AP-2 subunit [16, 17]. Alternatively, photoconversion experiments on cells expressing fluorescent protein-tagged clathrin or other appropriate adaptor molecules can reveal alterations in coated pit dynamics on intracellular membranes, such as endosomes or the trans Golgi complex [18]. Third, crystallographic studies should allow for the identification of binding site(s) on the clathrin TD which are targeted by the identified compounds [16]. Fourth, potential cytotoxic effects of all compounds displaying activity in vitro and in vivo should be assessed carefully. Finally, specificity tests should rule out that identified compounds interfere with other interactions within the endocytic network or other vesicle coats.

4 Notes

1. Alternatively, use 2× YT powder offered by some companies. Dissolve 31 g in 1 L of water. In this case pH does not need to be adjusted.

2. High-level expression of heterologous proteins can lead to a depletion of bacterial tRNAs and thereby stall translation. This strain contains extra copies of tRNAs that are rare in bacteria, but higher abundant in mammals.

3. Folding of eukaryotic proteins in prokaryotic systems can be inefficient for various reasons. Reducing temperature during protein expression can alleviate this problem in many cases.

4. Recover supernatant in appropriately sized container, and subject together with empty flasks to autoclaving.

5. Cell lysis releases proteases that can potentially cleave recombinant, nonbacterial proteins.

6. Many endocytic proteins contain polybasic stretches in their primary amino acid sequence. Thus, when expressed as recombinant proteins in bacteria, they are frequently found associated with negatively charged RNA and DNA. In in vitro binding assays associated nucleotides frequently mask protein regions involved in the interaction. This can be counter-acted by the addition of benzonase during the purification of these proteins, as benzonase is an endonuclease that degrades all forms of DNA and RNA, but has no proteolytic activity.

7. Keep cells on ice to prevent from heating up during the sonication procedure.

8. We use the Äkta explorer system (GE Healthcare) for protein purification by column chromatography, and usually pack columns on our own. The volume of column material depends on the expected protein yield. The material used here has a binding capacity of approximately 40 mg/mL. Alternatively, prepacked formats are available in various scales.

9. This step is required to remove precipitated protein or other particulate material that would clog the column in the subsequent chromatography step.

10. The GE Superdex 200 16/60 column will separate the sample according to size. Large proteins and oligomeric complexes will elute in early fractions, whereas small proteins will elute later. To calibrate the column a mixture of proteins of known sizes should be run. At the same time elution buffer is replaced by storage buffer.

11. This column material binds with high capacity (approximately 25 mg/mL) to GST-tagged fusion proteins.

12. Do not let the plate dry out from this point on. Make sure that the next solution is available before removing the current one from the wells.

13. This step serves to block residual binding sites in the well that have not been coupled to the His_6-tagged amphiphysin B/C domain. If blocking is inefficient, coupling of the GST-clathrin TD might occur and lead to false-negative results.

14. In case large numbers of compounds are to be tested a pipetting robot should be used to allow for accuracy and speed.

15. The effect of each compound should be evaluated in duplicate or triplicate samples.

16. As the screen will finally detect the amount of GST-tagged clathrin TD retained in each well, omission of the binding ligand will allow for the detection of the background signal.

17. The dilution of the antibody might have to be adapted when another supplier for the antibody is chosen, or when a new batch is used.

18. This well does not need to be coated with His_6-tagged amphiphysin B/C domain, and will not be washed.

19. Be cautious when using the acid-containing stop solution.

20. The Z′ factor serves as a measure to judge if the size of the response in the assay is large enough to allow for valid conclusions. It is defined by the means (μ) and standard deviations (σ) of the positive (pos: DMSO; row A in Fig. 3) and negative (neg: no GST-clathrin TD added; wells K1–K3) controls, according to the formula:

$$Z' = 1 - \left[3 \left(\sigma_{pos} + \sigma_{neg} \right) / \left| \left(\mu_{pos} - \mu_{neg} \right) \right| \right]$$

The Z′ factor should be between 0.5 and 1 to indicate that the screen produced data with sufficient sensitivity and robustness.

21. This dilution yields final concentrations of 300 μM, 100 μM, 30, μM, 10 μM, 3 μM, 1 μM, 0.3 μM, and 0.1 μM.

22. These experiments will also elucidate which of the identified compounds are membrane-permeable. Our initial study led to the identification of Pitstop1 and Pitstop2, both of which inhibit binding of the amphiphysin B/C domain in vitro, but only the latter of which blocks CME at low μM concentration because it can enter cells much more efficiently than the charged molecule Pitstop1.

23. The predicted molecular weight of the His_6-tagged amphiphysin B/C domain is 38.5 kD. Due to the enrichment of

negatively charged amino acids in this domain the apparent molecular weight is significantly higher.

24. The predicted molecular weight of the GST-tagged clathrin TD is 74.8 kD.

Acknowledgments

This work has been supported by grants from the Deutsche Forschungsgemeinschaft (SFB765/b4 to V.H. and SFB958/A11 to M.K.).

References

1. van der Bliek AM, Redelmeier TE, Damke H, Tisdale EJ, Meyerowitz EM, Schmid SL (1993) Mutations in human dynamin block an intermediate stage in coated vesicle formation. J Cell Biol 122:553–563

2. Evergren E, Marcucci M, Tomilin N, Low P, Slepnev V, Andersson F, Gad H, Brodin L, De Camilli P, Shupliakov O (2004) Amphiphysin is a component of clathrin coats formed during synaptic vesicle recycling at the lamprey giant synapse. Traffic 5:514–528

3. Slepnev VI, Ochoa GC, Butler MH, De Camilli P (2000) Tandem arrangement of the clathrin and AP-2 binding domains in amphiphysin 1 and disruption of clathrin coat function by amphiphysin fragments comprising these sites. J Biol Chem 275:17583–17589

4. Zhao X, Greener T, Al-Hasani H, Cushman SW, Eisenberg E, Greene LE (2001) Expression of auxilin or AP180 inhibits endocytosis by mislocalizing clathrin: evidence for formation of nascent pits containing AP1 or AP2 but not clathrin. J Cell Sci 114:353–365

5. Hinrichsen L, Harborth J, Andrees L, Weber K, Ungewickell EJ (2003) Effect of clathrin heavy chain- and alpha-adaptin-specific small inhibitory RNAs on endocytic accessory proteins and receptor trafficking in HeLa cells. J Biol Chem 278:45160–45170

6. Motley A, Bright NA, Seaman MN, Robinson MS (2003) Clathrin-mediated endocytosis in AP-2-depleted cells. J Cell Biol 162:909–918

7. Macia E, Ehrlich M, Massol R, Boucrot E, Brunner C, Kirchhausen T (2006) Dynasore, a cell-permeable inhibitor of dynamin. Dev Cell 10:839–850

8. Hill TA, Gordon CP, McGeachie AB, Venn-Brown B, Odell LR, Chau N, Quan A, Mariana A, Sakoff JA, Chircop M et al (2009) Inhibition of dynamin mediated endocytosis by the dynoles--synthesis and functional activity of a family of indoles. J Med Chem 52:3762–3773

9. McCluskey A, Daniel JA, Hadzic G, Chau N, Clayton EL, Mariana A, Whiting A, Gorgani NN, Lloyd J, Quan A et al (2013) Building a better dynasore: the dyngo compounds potently inhibit dynamin and endocytosis. Traffic 14:1272–1289

10. Park RJ, Shen H, Liu L, Liu X, Ferguson SM, De Camilli P (2013) Dynamin triple knockout cells reveal off target effects of commonly used dynamin inhibitors. J Cell Sci 126:5305–5312

11. Preta G, Cronin JG, Sheldon IM (2015) Dynasore - not just a dynamin inhibitor. Cell Commun Signal 13:24

12. Drake MT, Traub LM (2001) Interaction of two structurally distinct sequence types with the clathrin terminal domain beta-propeller. J Biol Chem 276:28700–28709

13. Kang DS, Kern RC, Puthenveedu MA, von Zastrow M, Williams JC, Benovic JL (2009) Structure of an arrestin2-clathrin complex reveals a novel clathrin binding domain that modulates receptor trafficking. J Biol Chem 284:29860–29872

14. Kern RC, Kang DS, Benovic JL (2009) Arrestin2/clathrin interaction is regulated by key N- and C-terminal regions in arrestin2. Biochemistry 48:7190–7200

15. Miele AE, Watson PJ, Evans PR, Traub LM, Owen DJ (2004) Two distinct interaction motifs in amphiphysin bind two independent sites on the clathrin terminal domain beta-propeller. Nat Struct Mol Biol 11:242–248

16. von Kleist L, Stahlschmidt W, Bulut H, Gromova K, Puchkov D, Robertson MJ, MacGregor KA, Tomilin N, Pechstein A, Chau N et al (2011) Role of the clathrin terminal

domain in regulating coated pit dynamics revealed by small molecule inhibition. Cell 146:471–484

17. Merrifield CJ, Feldman ME, Wan L, Almers W (2002) Imaging actin and dynamin recruitment during invagination of single clathrin-coated pits. Nat Cell Biol 4:691–698

18. Stahlschmidt W, Robertson MJ, Robinson PJ, McCluskey A, Haucke V (2014) Clathrin terminal domain-ligand interactions regulate sorting of mannose 6-phosphate receptors mediated by AP-1 and GGA adaptors. J Biol Chem 289:4906–4918

Chapter 6

Acute Manipulations of Clathrin-Mediated Endocytosis at Presynaptic Nerve Terminals

Rylie B. Walsh, Ona E. Bloom, and Jennifer R. Morgan

Abstract

Acute perturbations of clathrin and associated proteins at synapses have provided a wealth of knowledge on the molecular mechanisms underlying clathrin-mediated endocytosis (CME). The basic approach entails presynaptic microinjection of an inhibitory reagent targeted to the CME pathway, followed by a detailed ultrastructural analysis to identify how the perturbation affects the number and distribution of synaptic vesicles, plasma membrane, clathrin-coated pits, and clathrin-coated vesicles. This chapter describes the methodology for acutely perturbing CME at the lamprey giant reticulospinal synapse, a model vertebrate synapse that has been instrumental for identifying key protein–protein interactions that regulate CME in presynaptic nerve terminals with broader extension to nonneuronal cell types.

Key words AP2, Clathrin-coated vesicles, Electron microscopy, Lamprey, Synapse, Ultrastructure

1 Introduction

Neurotransmission depends critically upon the local recycling of synaptic vesicles at presynaptic nerve terminals. Following depolarization and calcium influx into the presynaptic nerve terminal, neurotransmitter-filled synaptic vesicles fuse with the presynaptic active zone and release their contents into the synaptic cleft (Fig. 1) [1]. Vesicular membrane must then be locally recycled from the areas surrounding the active zone. One of the primary mechanisms for recycling synaptic vesicles is clathrin-mediated endocytosis (CME), though clathrin-independent mechanisms may also participate [2, 3]. Briefly, clathrin coat formation is initiated when clathrin is recruited to the plasma membrane by the assembly proteins, AP180 and AP2 (Fig. 1). Clathrin coat assembly then promotes invagination and maturation of the budding vesicle, a process that is assisted by epsin, endophilin, and actin. Dynamin

The original version of this chapter was revised. A correction to this chapter can be found at
https://doi.org/10.1007/978-1-4939-8719-1_19

Laura E. Swan (ed.), *Clathrin-Mediated Endocytosis: Methods and Protocols*, Methods in Molecular Biology, vol. 1847,
https://doi.org/10.1007/978-1-4939-8719-1_6, © Springer Science+Business Media, LLC, part of Springer Nature 2018

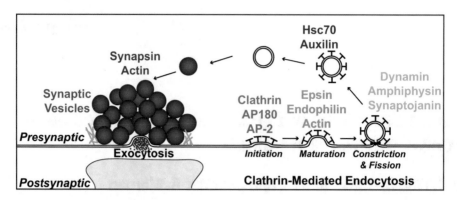

Fig. 1 Clathrin-mediated endocytosis at presynaptic nerve terminals. After exocytosis and neurotransmitter release, synaptic vesicles are locally recycled via CME. Shown here are the major transitions in CME and some of the proteins that participate in these transitions, which have been extensively studied using acute perturbations at synapses

and several effector proteins drive constriction at the neck of the clathrin-coated pit (CCP), and the GTPase activity of dynamin leads to fission and generation of a free clathrin-coated vesicle (CCV). Once the CCV is uncoated by the ATPase Hsc70 and its cochaperone, auxilin, the vesicle is then refilled with neurotransmitter and returned to the synaptic vesicle cluster. Synapsin and actin are involved in vesicle clustering. Because CME is triggered by neuronal activity at synapses, and this is under the experimenter's control, the neuronal synapse has proven to be a great cellular model for studying the molecular mechanisms of CME.

Electron microscopic (EM) studies of synapses within the frog neuromuscular junction, squid giant axon, and lamprey giant reticulospinal axon have been advantageous for studying the molecular components, morphological stages, and physiological correlates of CME [2, 4, 5]. These preparations allow for acute perturbations of CME and have provided a complementary approach to chronic genetic ablations or manipulations in other models. With acute perturbations, reagents such as peptides, recombinant proteins, antibodies, drugs, or toxins that are known or hypothesized to interfere with CME can be delivered via axonal microinjection directly to living, intact presynaptic terminals. Following neuronal stimulation to stimulate exocytosis and endocytosis at synapses, the tissue is fixed and processed for EM. Because clathrin coats are electron dense, EM can be used to determine the precise step in the clathrin pathway that is affected by the perturbation. Several advantages of acute manipulations include the opportunity to rapidly screen reagents for morphological effects and a lack of molecular compensation that may complicate data interpretation with chronic perturbations.

The giant reticulospinal (RS) synapses within the spinal cord of the lamprey (Figs. 2 and 3) have been particularly useful for elucidating molecular mechanisms of CME [6–13]. Giant RS synapses are large *en passant*, glutamatergic synapses (1–2 μm diameter)

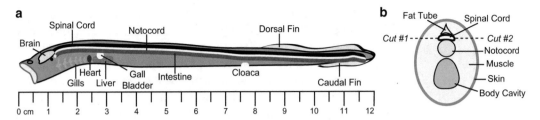

Fig. 2 Anatomy of the lamprey. (**a**) Diagram of a late larval lamprey showing the location of the spinal cord and other major organs. (**b**) Cross-section of a larval lamprey at the level of the gills. Note the position of the spinal cord. Dissection cuts for isolating the spinal cord are indicated

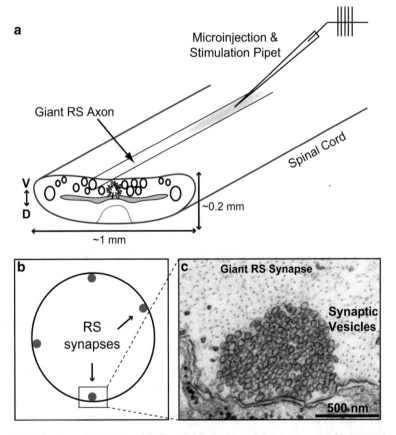

Fig. 3 Acute perturbations of CME at lamprey giant RS synapses. (**a**) Diagram of the lamprey spinal cord and axonal microinjection. *D* dorsal; *V* ventral. (**b**) Cross-section of a giant RS axon showing the synapses along the perimeter. (**c**) An electron micrograph of an unstimulated lamprey giant RS synapse showing a large pool of tightly clustered synaptic vesicles

located along the perimeter of giant axons (40–80 μm diameter), which are located within the ventromedial tract of the ribbon-like spinal cord (Fig. 3a–c). The large diameter and highly characteristic morphology of the giant RS axons and synapses, as well as the lack of myelin and relative transparency of the spinal cord, make the lamprey model particularly attractive for synaptic imaging, including ultrastructural analyses of the molecular mechanisms of CME. Large clusters of ~1000 synaptic vesicles are clustered at individual active zones (Fig. 3c), which are typically separated by fairly large distances, making it possible to attribute the synaptic vesicles and endocytic intermediates to a particular synapse. Since the RS synapses in the isolated spinal cord are quiescent, it is possible to trigger extended periods of exocytosis and endocytosis via action potentials that are evoked by electrical stimulation. Before, during, or after the stimulation period, perturbing reagents can be introduced to the synapses via axonal microinjection, and effects of such compounds on CME are subsequently determined.

Emphasizing its utility, acute perturbations at lamprey RS synapses have already been used successfully to identify the functions of many proteins with essential roles in CME and to define discrete stages of clathrin assembly, maturation, constriction/fission, and uncoating (Fig. 1) [5, 6, 8–14]. Complementary studies on CME have been done at the squid giant synapse [4, 15–18]. In both preparations, with low to moderate levels of activity, CME appears to be the predominant method for endocytosis [11, 15, 16]. This chapter describes how to acutely perturb CME at lamprey RS synapses, using microinjection of an AP2 peptide, which inhibits clathrin assembly in vitro and in vivo, as an example reagent. As previously demonstrated at the squid giant synapse [16], and as predicted for AP2's role in initiating clathrin assembly (Fig. 1), the AP2 peptide also causes a phenotype at lamprey RS synapses that is consistent with impairing very early stages of CME. We anticipate that acute perturbations at lamprey giant RS synapses will continue to be a useful approach for discovery-driven investigation into the molecular mechanisms of CME.

2   Materials

The following materials and methods have been adapted from previous references [6, 8–12, 14, 19].

2.1   Spinal Cord Dissections

1. Late stage larval sea lampreys (*Petromyzon marinus*; 10–13 cm). As lampreys are vertebrates, all procedures must be approved by the local institutional animal care and use committee prior to experimentation.

2. Anesthesia: MS-222 (0.1 g/L) buffered to pH 7.4 with sodium bicarbonate.

3. Sylgard® 184. Using the manufacturer's protocol, make ~100 mL of Sylgard mixture. Pour Sylgard into two large petri dishes (100 mm) to a thickness of 5–7 mm. With the remaining Sylgard, pour 5–10 small petri dishes (35 mm) to a thickness of 3–4 mm. Cure on a level surface at room temperature 1–3 days (*see* **Note 1**).

4. Lamprey Ringer's solution: 100 mM NaCl, 2.1 mM KCl, 2.6 mM CaCl$_2$, 1.8 mM MgCl$_2$, 4 mM glucose, 0.5 mM glutamine, 2 mM HEPES pH 7.4. Make up to 2 L. Adjust pH to 7.4 with NaOH. Store at 4 °C up to 2 weeks. Immediately prior to use, aerate for 20 min with oxygen.

5. Dissecting tools: two pairs of Dumont Dumoxel #5 forceps, Student Vannas spring scissors, standard or carbon steel razor blades, minutien pins. Cut 4–6 minutien pins in half, and place the pointed ends in the small Sylgard dish. Discard the blunt ends.

2.2 Buffers and Peptides

1. AP2 peptide: Peptides should be custom synthesized by commercial suppliers to >95% purity by HPLC (Quality and purity of peptide will have a significant influence on these experiments.). The amino acid sequence is shown in Table 1. A control AP2 peptide, in which the second DLL motif is mutated to AAA (e.g., mutant AP2 peptide), should also be synthesized [16]. Separate the lyophilized peptides into 1–2 mg aliquots and store at −80 °C until use.

2. Lamprey internal solution: 180 mM KCl, 10 mM HEPES, pH 7.4. Make up to 50 mL. Syringe filter into 0.5–1 mL aliquots. Store at −20 °C until use.

3. Alexa Fluor® 488 dextran (3000 MW, anionic) (*see* **Note 2**).

2.3 Microinjection and Electrophysiology Components

1. Premium standard wall borosilicate capillary glass with filament (*see* **Note 3**).

2. Electrode puller (e.g., P-97 or P-2000 from Sutter Instrument Company; Novato, CA)

3. 10 μL Glass Hamilton syringe with standard beveled 32G needle

4. Compound upright microscope with 10× air and 40× water dipping objectives (e.g., Zeiss Axioskop 2FS or similar)

5. Motorized micromanipulator with capabilities to drive fine movements in the nanometer to micrometer range and a diagonal axis, such as a Sutter MP-285 or Burleigh PCS-6000 (ThorLabs).

6. Straight electrode holder with side port for attaching tubing from pressure injector (*see* **item 9**).

7. Intracellular amplifier, such as Axon Instruments Axoclamp 2B or Molecular Devices Multiclamp 700B. Mount an amplifier headstage (1× or 10×) onto the motorized micromanipulator, and plug in electrode holder from step 6. Plug headstage into the amplifier.

8. Computer or oscilloscope: Connect the output of the correct amplifier channel from Step 7 into a computer running pClamp

Table 1
List of reagents used to acutely perturb CME at synapses and their effects. Details of their effects on synaptic morphology and function, as well as control reagents, are described in the references

Stage of CME affected	Reagent	Biochemical actions	Sequence	References
Clathrin assembly	AP2 peptide	Binds clathrin N-terminal domain and inhibits clathrin assembly in vitro and in vivo	QGDLLGDLLNLDLGPPVNVPQ	[16, 20]; Fig. 3
	AP180 peptide	Inhibits clathrin assembly in vitro and in vivo	SGGATAWGDLLGEDSLAALSS	[16]
	Pitstop 1	Inhibits ligand interactions with clathrin N-terminal domain	n.a.	[13]
Clathrin coat maturation	Endophilin antibody	Inhibits CME and vesicle recycling in vivo; build up of shallow CCPs	Antibody raised against a peptide corresponding to a.a. 255–274 of rat endophilin (YQPKPRMSLEFATGDGTQPN)	[10]
	Actin toxins (Phalloidin, Bot C2; Lat B, Swinholide)	Inhibit actin dynamics in vivo	n.a.	[6, 12]
	PIP kinase peptide	Inhibits actin polymerization in vivo	TDERSWVYSPLHYS	[9]
Constriction and fission	Dynamin peptide	Inhibits dynamin-amphiphysin SH3 interaction	PPPQVPSRPNRAPPG	[11]
	Amphiphysin SH3 domain	Binds dynamin in vitro and prevents recruitment to CCP	GST-tagged SH3 domain of human amphiphysin (a.a. 623–694)	[11]
Fission and uncoating	Synaptojanin peptide	Inhibits endophilin interactions with dynamin and synaptojanin in vitro	VAPPARPAPPQRPPPSGA	[8]
	Auxilin ΔHPD mutant	Inhibits Hsc70-auxilin interaction and clathrin uncoating in vitro	Recombinant bovine auxilin truncation (a.a. 547–910) with mutated HPD motif (a.a. 874–876)	[17]
	α-Synuclein	Binds Hsc70	Recombinant full-length human α-synuclein (a.a. 1–140)	[21]

software (or equivalent), or alternatively into an oscilloscope so that the neuronal signals can be monitored during injections and stimulation.

9. Pressure injector: Picospritzer III or similar. Set up the injector according to manufacturer's instructions. The small diameter tubing should be firmly sealed onto the side port of the electrode holder (*see* **Note 4**).

2.4 Electron Microscopy

Make all solutions in a fume hood and follow appropriate safety precautions.

1. Cacodylate Buffer: Stock solution of 200 mL of 0.2 M sodium cacodylate, pH 7.4. Dilute 1:1 with distilled water to make 100 mL of 0.1 M Na cacodylate for buffer washes.

2. Glutaraldehyde fixative: Make 3% glutaraldehyde–2% paraformaldehyde in 0.1 M sodium cacodylate buffer, pH 7.4. Heat 50 mL of 0.2 M sodium cacodylate buffer, pH 7.4, to 70 °C, and stir on a heating stirplate. Add a few (~4) drops of 5 N NaOH to increase the pH. Add 2 g of paraformaldehyde prills. Stir until paraformaldehyde is completely in solution. pH to 7.4. Adjust volume to 70 mL with distilled water. Filter 23.3 mL of the paraformaldehyde fixative into a 50 mL conical tube using a 0.22 μm syringe filter. Add 10 mL of 10% aqueous glutaraldehyde using a glass Pasteur pipette (*see* **Note 5**). Confirm that pH is still at 7.4. Prepare on the same day as the experiments.

3. Osmium fixative: Use extreme caution! Immediately prior to post-fixing (Subheading 3.3, **step 3**), prepare 4 mL of 2% osmium tetroxide/2% K^+ ferrocyanide in 0.1 M sodium cacodylate, pH 7.4. Dissolve 0.2 g K+ ferrocyanide in 5 mL of 0.2 M Na cacodylate, pH 7.4. Mix 2 mL of this solution with a 2 mL ampoule of 4% aqueous OsO_4. Store solution in a designated refrigerator only used for osmium tetroxide.

4. Glass scintillation vials—20 mL capacity, with caps.

5. 2 bottles of 100% ethanol. Using one bottle, prepare 50–100 mL stocks of 50%, 70%, and 95%.

6. Propylene oxide, EM grade.

7. Embedding supplies: EMbed-812 resin kit, Wooden sticks, Dental wax sheets; flat embedding mold; dry oven set to 60 °C.

8. Ultramicrotome, such as Leica EM UC7 or Reichert-Jung Ultracut.

9. Glass knife maker, such as Leica EM KMR3 or similar. Make 4–6 glass knives, and store in a lint-free environment until use. This video shows how to make glass knives: https://www.youtube.com/watch?v=NJcLhM3sfEM. The more experienced EM microscopist will use a diamond knife for thin sectioning.

10. Sectioning supplies: glass knife boats; hot pen; formvar-coated copper slot grids; grid mats.

11. 1% toluidine blue: Mix 1 g borax and 1 g toluidine blue in 100 mL ddH$_2$O water, and stir overnight.

12. Counterstaining solutions: Uranyl acetate: Make 50 mL of a 2% solution. Store at 4 °C in the dark until use. Lead citrate: Make 50 mL of a 0.4% solution. Store at 4 °C until use. Make 50 mL of 0.2 N NaOH, and store at room temperature.

13. Electron microscope, such as an FEI Technai Spirit BioTwin T12.

3   Methods

3.1   Spinal Cord Dissection

1. Anesthetize a lamprey in buffered 0.1 g/L tricaine methane-sulfonate diluted in tank water. Complete anesthesia takes ~10–15 min and is determined by lack of spontaneous muscle contractions, no response to tail pinch, and slowed gill contractions.

2. After anesthesia is complete, move the lamprey to a paper towel. Using a razor blade, decapitate near the level of the second gill (Fig. 2a) and pith. Make a second cut at the level of the dorsal fin to remove the tail.

3. Move the lamprey body piece to one of the large Sylgard dishes. Pin on both ends using syringe needles, and submerge in fresh, oxygenated lamprey Ringer's solution.

4. Using dissecting scissors, make two horizontal cuts through the cartilaginous casing at the lateral edges of the spinal cord (Fig. 2b). Extend the cuts through the muscle and skin on both sides of the animal. As these two cuts are extended rostrally, the top portion of the body can be lifted up with forceps to reveal the dorsal surface of the spinal cord. Continue in this manner until the entire spinal cord is revealed (*see* **Note 6**).

5. Isolate a piece (2–3 cm) of the spinal cord. Hold onto one end of the spinal cord with the forceps, and gently peel it up out of the cartilage encasing until it is free on both ends (*see* **Note 7**).

6. Move the spinal cord to a small Sylgard dish and submerge in fresh, oxygenated lamprey Ringer's solution. Pin the spinal cord ventral side up by placing 1–2 minutien pins in each end of the spinal cord (*see* **Note 8**).

7. Using fine forceps, remove the thin layer of *meninx primativa* from the ventral surface of the spinal cord (*see* **Note 9**).

8. Store spinal cord in lamprey Ringer's solution at 4 °C while the peptides are being prepared.

3.2 Preparation of Peptides and Microinjection

1. Pull 5–10 sharp microelectrodes using the electrode puller (*see* **Note 10**). These will be used for peptide microinjections and for stimulating the axons.

2. Dilute an aliquot of the AP2 peptide (or control peptide) in lamprey internal solution to a concentration of 20 mM or higher (*see* **Note 11**). pH to 7.4 using 5 M KOH. The final volume of the peptide aliquot typically ranges from 10–30 μL depending on the molecular weight of the peptide.

3. Add Alexa Fluor® 488 dextran to the peptide aliquot to a final dye concentration of 100 μM. Mix by pipetting up and down.

4. Centrifuge the peptide at $15,700 \times g$ in a tabletop microfuge for 10 min at 4 °C to remove any aggregates.

5. Immediately load 1–2 μL of the peptide solution into several microelectrodes using the Hamilton syringe. Store the remaining peptide solution on ice. Mount a peptide-containing microelectrode onto the electrode holder and secure it tightly.

6. Place the small dish containing the lamprey spinal cord onto the stage of an appropriate upright microscope, making sure it is securely fastened. Place a minutien pin in or beside the lateral edge of the spinal cord, away from the area of the large RS axons, in a position parallel to the injection site. The pin will remain in place throughout the fixation and will later serve as a landmark to identify the injection site during the EM analysis.

7. Using the coarse movements on the micromanipulator, position the tip of the microelectrode just above the axon to be injected. Make sure that the microelectrode is oriented along the same axis as the selected axon (i.e., parallel with the longitudinal axis of the spinal cord), as this will allow for easier axon impalements (Fig. 3a).

8. Switch the micromanipulator to the fine movements, and slowly advance the microelectrode until the tip just enters the axon, registering a membrane potential of at least −58 mV (*see* **Note 12**). Wait 1–2 min to ensure that the resting membrane potential is stable before injecting.

9. Inject the peptide using brief pulses of N_2 delivered through the pressure injector (5–40 ms; 0.1–0.3 Hz; 30–50 psi). Start with the lowest settings, and then increase as needed until each pulse produces a small burst of fluorescence in the axon.

10. Continue injecting pulses of peptide for 5–30 min, until the desired concentration is achieved (*see* **Note 13**).

11. While continuing to inject, stimulate the axon at the desired rate by delivering current pulses (1 ms; 20–80 nA; 0.5–20 Hz) to evoke action potentials (*see* **Note 14**). Start by injecting a 1 nA current, and then increase the amplitude until the axon

reliably fires action potentials. The most common stimulation paradigms for eliciting CME at lamprey synapses are 5 Hz for 30 min [8, 10, 11, 22] or 20 Hz for 5 min [6, 9, 14, 22], though the latter may also evoke bulk endocytosis [22].

12. At the end of the stimulation period, while continuing to stimulate, carefully remove the lamprey Ringer's solution from the chamber and replace with freshly prepared glutaraldehyde fixative. Triturate 10–20 times with a disposable transfer pipette. The action potentials will disappear within 20–30 sec, as the tissue fixes. Gently remove the microelectrode. Immediately transfer the fixed spinal cord to a fume hood. Wash 3 × 5 min with fresh fixative, triturating each time. Continue fixing the spinal cords for 2–5 h at room temperature, replacing with fresh fixative approximately once per hour. Store overnight at 4 °C (*see* **Note 15**).

13. Repeat steps 5–13 until the desired number of preparations has been fixed.

3.3 Fixation and Processing for Electron Microscopy

The general steps for EM processing are: fixation, dehydration, resin infiltration, and curing of preparation in resin molds. Perform all incubations in a fume hood and wear appropriate personal protective gear.

1. Remove the pins from the spinal cord, and transfer it to a scintillation vial containing 0.1 M sodium cacodylate buffer, pH 7.4.

2. Wash 3 × 10 min in 0.1 M sodium cacodylate buffer.

3. Postfix for 1.5 hours in freshly prepared K^+ ferrocyanide/OsO_4 on ice and in the dark (*see* **Note 16**).

4. Wash 3 × 15 min in 0.1 M sodium cacodylate buffer, pH 7.4

5. Wash 3 × 15 min in ddH_2O water.

6. Perform *en bloc* staining with 2% aqueous uranyl acetate for 2 hours in the dark at room temperature (*see* **Note 17**).

7. Wash 3 × 5 min in ddH_2O water.

8. Dehydrate spinal cord in graded ethanol series, while rotating, as follows: 50% for 10 min; 70% for 15 min; 95% for 2 × 10 min; 100% for 2 × 15 min. Then, wash 2 × 30 min longer in 100% ethanol (EM grade) from a freshly opened bottle (or absolute EtOH stored over molecular sieves).

9. Wash in propylene oxide 2 × 15 min in a scintillation vial with the cap on (*see* **Note 18**).

10. Infiltrate the spinal cord in a 1:1 mixture of propylene oxide: EMbed 812 rotating overnight in the fume hood. Leave the cap off of the scintillation vial.

11. Move the spinal cord to a new scintillation vial, and cover with fresh EMbed 812. Infiltrate the preparation with EMbed 812 at room temperature for 4–5 hours.

12. During **step 11**, print small computer labels with the experiment date and condition, and place in the appropriate EM molds.

13. At the end of the infiltration, use a wooden stick to move the fixed spinal cord onto a piece of dental wax. Under a dissecting microscope, trim the spinal cord to a length of 2–3 mm, centered around the injection site, as identified by the pin mark (see 3.2.6). Place one end of the spinal cord flush with the tapered end of the embedding block, and cover with fresh EMbed 812 until the mold is filled.

14. Polymerize the blocks in a dry oven at 60 °C for >48 h until completely hardened.

3.4 Electron Microscopy and Image Analysis

1. Mount the spinal cord block in an ultramicrotome chuck. Using a fresh razor blade, trim the block face into a trapezoid shape around the spinal cord, keeping the longest edge at a length of ~1 mm.

2. Using a glass knife, with a ddH$_2$O-filled boat and mounted on the ultramicrotome, cut thick sections (1 μm) until the entire spinal cord fills the section. To determine this, collect several sections onto a microscope slide, stain the section with 1–2% toluidine blue for 2–3 min on a hot plate, destain by gently washing the sections with water, and visualize on a standard upright compound microscope (*see* **Note 19**). The toluidine blue-stained sections are used to assess the quality of the tissue after fixation and for identifying the target injected axons.

3. When within 200 μm of the injection site, start collecting ultrathin, silver sections (~70 nm), which can be obtained using a glass or diamond knife.

4. Collect 4–8 ultrathin serial sections, and place them on a formvar-coated copper slot grid. Use a hot pen to smooth out the sections before moving them onto the grid. Continue collecting sections at 5–20 μm intervals up to the injection site (marked by a large hole in the lateral edge of the spinal cord), until you have the desired number of grids (*see* **Note 20**).

5. Counterstain the sections: Using a syringe filter, place one drop of each solution onto a piece of parafilm for each grid. Grids can be moved between droplets using a wire loop or fine-pointed forceps. First, wash the grids by placing them on ddH$_2$O droplets for 1 min. To do so, place the grids section-side-down onto the drops of ddH$_2$O. Following the same procedure, stain the sections with 2% uranyl acetate for 7 min in the dark. Wash grids for 5 × 1 min in ddH$_2$O. Stain sections

with 0.4% lead citrate for 4 min (*see* **Note 21**). Wash grids in 0.2 N NaOH for 2 min. Finally, wash the grids in ddH$_2$O for 1 min. Carefully blot the edge of the grids dry with a tissue and place them to dry on a grid mat.

6. Load a grid into the specimen holder of the electron microscope. Identify the injected, stimulated axon, and acquire images of all synapses around the perimeter of that axon at the desired magnification (*see* **Note 22**). We typically use 26,500–37,000× magnification. Repeat until every section of every synapse from every grid has been imaged.

7. Using Image J or similar image analysis software, perform quantitative analyses on the synapses for your features of interest (*see* **Note 23**). First, select a section at or near the center of the synapse, as determined by the longest length of the active zone. Next, count the number of synaptic vesicles within a 1 μm distance from the active zone. Measure the size of the plasma membrane evaginations, as described in [22]. Briefly, draw a straight line (1 μm) from the edge of the active zone to the nearest point on the axolemma. Then, measure the curved distance between the two points. Repeat for the other side of the synapse, and determine the average. Next, identify the clathrin-coated pits (CCPs) and clathrin-coated vesicles (CCVs) by the appearance of an electron dense fuzzy coat (*see* Fig. 4g–i), and stage them accordingly. Stages are as follows: Stage 1—CCP with little invagination; Stage 2—invaginated, unconstricted CCP; Stage 3—invaginated, constricted CCP, sometimes exhibiting a dense ring of protein (e.g., dynamin); Stage 4—free CCVs [6, 14, 22].

8. Plot the average data from all synapses, and select representative images that show the resulting phenotype (Fig. 4a–f). Compared to the mutant AP2 peptide (control), the AP2 peptide significantly reduced the number of synaptic vesicles (Mut pep: 134 ± 30 SVs, $n = 10$ synapses; AP2 pep: 83 ± 10 SVs, $n = 20$ synapses; T-test, $p = 0.05$) (Fig. 4a–c). The AP2 peptide also increased the size of the plasma membrane evaginations (Mut pep: 2.0 ± 0.1 μm, $n = 10$ synapses; AP2 pep: 2.6 ± 0.2 μm, $n = 20$ synapses; T-test, $p = 0.05$) (Fig. 4d). However, no change was observed in the total number of clathrin-coated pits and vesicles (Mut pep: 1.5 ± 0.4 CCP/Vs, $n = 10$ synapses; AP2 pep: 1.9 ± 0.3 CCP/Vs, $n = 20$ synapses; T-test, $p = 0.48$)

Fig.4 (continued) Phenotypes observed when later stages of CME are acutely perturbed. As examples, disruption of presynaptic actin cytoskeleton, using a PIP kinase peptide, causes a large increase in CCPs that are invaginated, but unconstricted (**g**), while disruption of dynamin inhibits fission and increases the number of constricted CCPs (**h**). An interesting new finding is that excess human α-synuclein increases CCVs (**i**), suggesting effects on uncoating [21]. Scale bars = 0.2 μm. Panel **h** from [11]. Reprinted with permission from AAAS

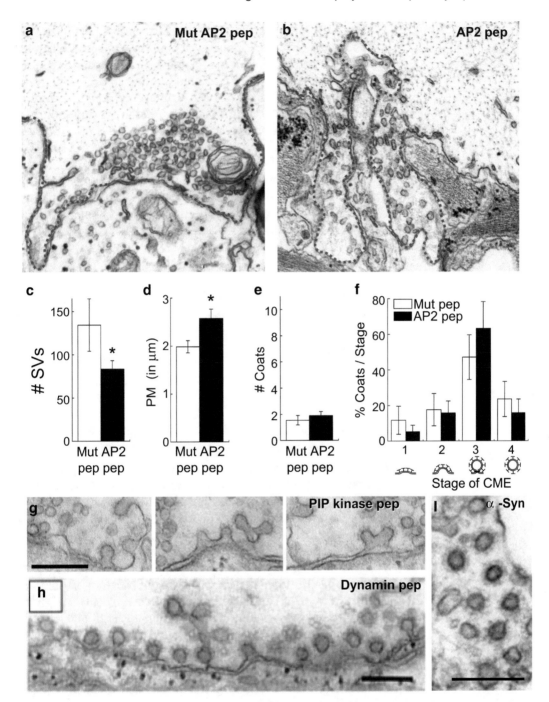

Fig. 4 Acute perturbations of CME at lamprey RS synapses with an AP2 peptide. (**a–b**) Electron micrographs of stimulated synapses (20 Hz, 5 min) treated either with a mutant AP2 peptide (control) or the AP2 peptide. Note the smaller synaptic vesicle cluster and larger plasma membrane evaginations (dotted line) with AP2 peptide, which is a characteristic hallmark of inhibiting CME at synapses. (**c–f**) Quantification of the phenotype produced by the AP2 peptide. AP2 peptide causes a loss of synaptic vesicles (**c**), increased plasma membrane evaginations (**d**), but no change in either the number or distribution of CCPs or CCVs (**e–f**). Bars represent mean ± SEM from **n** = 10–20 synapses; Asterisks indicate significance (**T**-test in **c–e**; ANOVA in **f**). (**g–i**)

(Fig. 4e). The distribution of CCPs and CCVs was also unchanged (Fig. 4f). Similarly, at lamprey and squid synapses, a loss of synaptic vesicles compensated by an expanded plasma membrane has also been observed in other studies where various stages of CME have been acutely perturbed, including clathrin assembly, clathrin coat maturation, constriction and fission, and clathrin uncoating [6, 8, 10, 13–17]. However, one major difference is that reagents that affect later stages of CME cause a >10-fold increase in the number of CCPs and CCVs at synapses. For example, actin-disrupting reagents impair clathrin coat maturation and cause a build-up of Stage 2 CCPs, while reagents that perturb dynamin function inhibit vesicle fission and cause a buildup of Stage 3 CCPs (Fig. 4g–h) [6, 9, 11, 12, 14]. Reagents that affect uncoating induce a build-up of Stage 4 CCVs (Fig. 4i) [17]. Table 1 is a nonexhaustive list of some of the reagents that perturb CME at lamprey giant RS synapses. Note the diversity of reagents, including peptides, toxins, recombinant proteins and antibodies, illustrating the many ways that CME can be acutely manipulated at synapses.

4   Notes

1. Sylgard dishes are reusable, even after exposed to fixative, as long as they are washed thoroughly between experiments.

2. A fluorescent dextran is coinjected along with the peptide in order to track the diffusion and concentration of the peptide throughout the axon. The 3000 MW fluorescent dextran provides a good proxy for most peptides, which have relatively low molecular weights. However, peptides may also be synthesized with a covalently attached fluorophore, such as FITC, for direct tracking of the peptide diffusion. For larger proteins or antibodies, a higher molecular weight dextran should be injected, one that is closely matched to the molecular weight of the protein. Alternatively, the protein or antibody can be directly conjugated with a fluorescent dye using an Alexa Fluor® labeling kit from Thermo Fisher, Inc.

3. The outer diameter (OD) of the capillary glass should be matched to the size of the port on the microelectrode holder, typically 1.0 or 1.2 mm.

4. Make sure the seal is tight so that each puff of air escapes only from the front end of the electrode holder. Be sure to secure any loose tubing, so that there is little to no movement with each puff of air. This will significantly lower the risk of mechanical damage to the axon during the microinjections.

5. Equilibrate the 10% aqueous glutaraldehyde to room temperature prior to use. Otherwise, it may precipitate in the tissue. For best results, the 3% glutaraldehyde–2% paraformaldehyde fixative should be made on the day of the experiment.

6. During the dissection, keep the scissors as horizontal as possible, and make small cuts. This will help to prevent nicking the spinal cord. In some cases, a thick layer of spongy connective tissue can be seen on top of the spinal cord, and if present it should be removed.

7. After removing the experimental piece of spinal cord, the rest of the animal can be stored in lamprey Ringer's at 4 °C up to 24 h. If the dissection went well, there should be enough remaining spinal cord for 1–2 additional experiments.

8. Stretch the spinal cord to the same length it was in the animal. Do not use the blunt ends of the minutien pins for pinning, as these create large holes that can tear or otherwise damage the tissue.

9. It is essential to remove the *meninx primativa* because the microinjection pipettes cannot easily penetrate it. Starting at one end of the spinal cord, use the fine forceps to pick at the top layer of the tissue until you can pull up a thin, translucent sheet of tissue. This is the *meninx.* If done carefully, the meninx can be removed in a single sheet from the entire ventral surface of the spinal cord. At the very least, remove the meninx from the middle portion of the spinal cord where the giant axons are located.

10. The Pipette Cookbook 2015 published by Sutter Instruments is an excellent resource for pulling sharp electrodes (http://www.sutter.com/PDFs/pipette_cookbook.pdf).See the chapter on Intracellular Recording Electrodes for specific settings.

11. Once prepared, the aliquot of peptide can be stored at 4 °C and used for experiments over 1–2 days. During the injection, the peptide is typically diluted 5–10 times from the stock concentration. Thus, the final axonal concentration of AP2 pep will be 2–4 mM, a range that inhibits synaptic transmission and synaptic vesicle recycling at squid giant synapses [16].

12. The best approach for impalement is to intersperse small microelectrode advances with short current pulses (1 ms) delivered through the "buzz" function on the amplifier. Upon entry of the microelectrode, a healthy lamprey giant axon will have a resting potential of −60 to −70 mV. The membrane potential may become slightly more hyperpolarized as the KCl in the electrode leaks out. However, if the membrane potential depolarizes, then the axon has not sealed properly and therefore is unlikely to be healthy enough to sustain the injection. Should this occur, discontinue and start over with a fresh piece of spinal cord.

13. The fluorescent dextran is used to estimate axonal concentration and diffusion of the peptide. To do so, compare the fluorescence intensity in the axon to a set of preprepared standard dilutions of the peptide solution. If the pipette clogs before the desired concentration is reached, try clearing the pipet by injecting a small pulse of current using the "buzz" or "clear" function on the amplifier. If this does not work, then gently remove the microelectrode and reimpale with a new one, being careful not to damage the axon.

14. Some peptide solutions do not allow enough current to pass through the electrode to spike action potentials. If that occurs, gently remove the peptide-containing microelectrode and replace it with one containing 3 M KCl for the stimulation.

15. The spinal cord can be stored in glut/para fixative at 4 °C for several days to weeks.

16. Osmium is extremely hazardous, and the vapors can fix mucous membranes of the eyes and nose and cause problems with breathing. Always work with osmium in a fume hood, and dispose of properly.

17. Uranyl acetate should be filtered at the time of use with a 0.22 μm syringe filter. UA is considered to be radioactive, but can be shielded by plastic, glass, or 6 in of distance, and gloves. Store and dispose of in a proper container.

18. Propylene oxide (PO) vapors are toxic, so work in the fume hood. Always use PO in glass or metal containers, as it can leech or dissolve plastics.

19. The giant axons should be visible, round, and uniformly stained within the ventromedial tract ([22] shows a good example). If the cytoplasm is pulled away from the axolemma, this is a clear indication that the fixation was suboptimal, and the preparation should be discontinued for EM.

20. It is important to keep track of where you are sectioning relative to the injection site. This will allow you to estimate the peptide concentration at a particular location in the axon and match it with the observed phenotype.

21. The length of the lead citrate staining can be altered to achieve the desired tissue contrast. Also, the lead citrate will form large precipitates on the sections if exposed to oxygen. To prevent this, perform the lead citrate staining in a glass petri dish filled with pellets of NaOH (which absorb CO_2) and sealed with parafilm. Do not breathe on the grids. Do not use lead citrate solution if it looks cloudy.

22. Synapses within the stimulated axon will exhibit a few clathrin-coated pits and vesicles, which are rarely if ever seen at unstimulated synapses. In addition, the vesicle cluster is slightly

smaller and more dispersed, and the plasma membrane is somewhat evaginated (*see* Fig. 3c and compared to Fig. 4a). A single synapse may span ten sections. It is useful to image all of the sections of a given synapse in the event that you want to perform 3D reconstructions from serial sections later on.

23. Do not perform image analysis on synapses within 25 μm of the injection site, because there can be local injection artifacts that could affect the results.

Acknowledgments

This work was supported by NIH grants: NINDS/NIA R01 NS078165 (to J.R.M.) and NIGMS R01 GM118933 (to E.M.L.). The authors would like to thank Dr. Eileen M. Lafer (University of TX Health Sciences Center, San Antonio) for providing the AP-2 peptide, as well as Paul Oliphint and Kara Marshall for technical assistance.

References

1. Pang ZP, Sudhof TC (2010) Cell biology of Ca2+−triggered exocytosis. Curr Opin Cell Biol 22:496–505

2. Heuser JE, Reese TS (1973) Evidence for recycling of synaptic vesicle membrane during transmitter release at the frog neuromuscular junction. J Cell Biol 57:315–344

3. Saheki Y, De Camilli P (2012) Synaptic vesicle endocytosis. Cold Spring Harb Perspect Biol 4:a005645

4. Augustine GJ, Morgan JR, Villalba-Galea CA, Jin S, Prasad K, Lafer EM (2006) Clathrin and synaptic vesicle endocytosis: studies at the squid giant synapse. Biochem Soc Trans 34:68–72

5. Brodin L, Shupliakov O (2006) Giant reticulospinal synapse in lamprey: molecular links between active and periactive zones. Cell Tissue Res 326:301–310

6. Bourne J, Morgan JR, Pieribone VA (2006) Actin polymerization regulates clathrin coat maturation during early stages of synaptic vesicle recycling at lamprey synapses. J Comp Neurol 497:600–609

7. Brodin L, Low P, Shupliakov O (2000) Sequential steps in clathrin-mediated synaptic vesicle endocytosis. Curr Opin Neurobiol 10:312–320

8. Gad H, Ringstad N, Low P, Kjaerulff O, Gustafsson J, Wenk M, Di Paolo G, Nemoto Y, Crun J, Ellisman MH, De Camilli P, Shupliakov O, Brodin L (2000) Fission and uncoating of synaptic clathrin-coated vesicles are perturbed by disruption of interactions with the SH3 domain of endophilin. Neuron 27:301–312

9. Morgan JR, Di Paolo G, Werner H, Shchedrina VA, Pypaert M, Pieribone VA, De Camilli P (2004) A role for Talin in presynaptic function. J Cell Biol 167:43–50

10. Ringstad N, Gad H, Low P, Di Paolo G, Brodin L, Shupliakov O, De Camilli P (1999) Endophilin/SH3p4 is required for the transition from early to late stages in clathrin-mediated synaptic vesicle endocytosis. Neuron 24:143–154

11. Shupliakov O, Low P, Grabs D, Gad H, Chen H, David C, Takei K, De Camilli P, Brodin L (1997) Synaptic vesicle endocytosis impaired by disruption of dynamin-SH3 domain interactions. Science 276:259–263

12. Shupliakov O, Bloom O, Gustafsson JS, Kjaerulff O, Low P, Tomilin N, Pieribone VA, Greengard P, Brodin L (2002) Impaired recycling of synaptic vesicles after acute perturbation of the presynaptic actin cytoskeleton. Proc Natl Acad Sci U S A 99:14476–14481

13. von Kleist L, Stahlschmidt W, Bulut H, Gromova K, Puchkov D, Robertson MJ, MacGregor KA, Tomilin N, Pechstein A, Chau N, Chircop M, Sakoff J, von Kries JP, Saenger W, Krausslich HG, Shupliakov O, Robinson PJ, McCluskey A, Haucke V (2011) Role of the clathrin terminal domain in regulating coated pit dynamics revealed by small molecule inhibition. Cell 146:471–484

14. Morgan JR, Jiang J, Oliphint PA, Jin S, Gimenez LE, Busch DJ, Foldes AE, Zhuo Y, Sousa R, Lafer EM (2013) A role for an Hsp70 nucleotide exchange factor in the regulation of synaptic vesicle endocytosis. J Neurosci 33:8009–8021

15. Morgan JR, Zhao X, Womack M, Prasad K, Augustine GJ, Lafer EM (1999) A role for the clathrin assembly domain of AP180 in synaptic vesicle endocytosis. J Neurosci 19:10201–10212

16. Morgan JR, Prasad K, Hao W, Augustine GJ, Lafer EM (2000) A conserved clathrin assembly motif essential for synaptic vesicle endocytosis. J Neurosci 20:8667–8676

17. Morgan JR, Prasad K, Jin S, Augustine GJ, Lafer EM (2001) Uncoating of clathrin-coated vesicles in presynaptic terminals: roles for Hsc70 and auxilin. Neuron 32:289–300

18. Morgan JR, Prasad K, Jin S, Augustine GJ, Lafer EM (2003) Eps15 homology domain-NPF motif interactions regulate clathrin coat assembly during synaptic vesicle recycling. J Biol Chem 278:33583–33592

19. Pieribone VA, Shupliakov O, Brodin L, Hilfiker-Rothenfluh S, Czernik AJ, Greengard P (1995) Distinct pools of synaptic vesicles in neurotransmitter release. Nature 375:493–497

20. Zhuo Y, Cano KE, Wang L, Ilangovan U, Hinck AP, Sousa R, Lafer EM (2015) Nuclear magnetic resonance structural mapping reveals promiscuous interactions between Clathrin-box motif sequences and the N-terminal domain of the Clathrin heavy chain. Biochemistry 54:2571–2580

21. Banks SM, Busch DJ, Oliphint PA, Walsh RB, George JM, Lafer EM, Morgan JR (2015) α-Synuclein interacts with Hsc70: a possible mechanism underlying the synaptic vesicle recycling defects in Parkinson's Disease models. Soc Neurosci Abst 36.13, Chicago, IL

22. Busch DJ, Oliphint PA, Walsh RB, Banks SM, Woods WS, George JM, Morgan JR (2014) Acute increase of alpha-synuclein inhibits synaptic vesicle recycling evoked during intense stimulation. Mol Biol Cell 25:3926–3941

Chapter 7

Imaging "Hot-Wired" Clathrin-Mediated Endocytosis

Laura A. Wood and Stephen J. Royle

Abstract

Clathrin-mediated endocytosis (CME) occurs continuously at the plasma membrane of eukaryotic cells. However, when a vesicle forms and what cargo it contains are unpredictable. We recently developed a system to trigger CME on-demand. This means that we can control when endocytosis is triggered and the design means that the cargo that is internalized is predetermined. The method is called hot-wired CME because several steps and proteins are bypassed in our system. In this chapter, we describe in detail how to use the hot-wiring system to trigger endocytosis in human cell lines and how to image the vesicles that form using microscopy and finally, how to analyze those images.

Key words Clathrin, Endocytosis, Chemically induced dimerization, Live cell imaging, Immunofluorescence, Image analysis

1 Introduction

Clathrin-mediated endocytosis (CME) is a high capacity pathway for uptake of a variety of transmembrane proteins into eukaryotic cells. Constitutive CME occurs continuously at the cell surface but the initiation of each event is unpredictable. Both the timing and the contents of the vesicle are difficult to foresee [1, 2]. Previous work in vitro described how clathrin could be recruited to nucleate new pits using a clathrin-binding protein fragment attached to a planar membrane [3]. We recently expanded this work to develop a system to trigger endocytosis on-demand in living human cells [4].

During constitutive CME, a pit is initiated by a complex multistep process which in part requires the adaptor protein complex AP2 to undergo several conformational changes before it is able to engage cargo and membrane, and then recruit clathrin [5]. The system we developed removed a number of these intermediary steps such that we send a clathrin-binding protein (clathrin

The original version of this chapter was revised. A correction to this chapter can be found at https://doi.org/10.1007/978-1-4939-8719-1_19

Laura E. Swan (ed.), *Clathrin-Mediated Endocytosis: Methods and Protocols*, Methods in Molecular Biology, vol. 1847, https://doi.org/10.1007/978-1-4939-8719-1_7, © Springer Science+Business Media, LLC, part of Springer Nature 2018

"hook") to a plasma membrane "anchor" to mimic the final stage of pit initiation. Because we have bypassed a number of steps we refer to this process of triggering endocytosis on-demand as "hot-wired" CME.

The hook and anchor are brought together by the heterodimerization of FKBP and FRB domains using rapamycin [6]. This means that both the timing of initiation and the content of the vesicle are controlled. In the basic design, the plasma membrane anchor is the α chain of CD8 to which good antibodies can be used to track internalization [7]. However, other transmembrane and even peripheral membrane proteins such as CD4, Fyn, and GAP43 can also be used. For the clathrin hook we use a fragment of the β2 subunit from the AP2 complex (hinge and appendage regions). Other hooks can be used, such as the hinge and appendage region from the β1 subunit from AP1. As a negative control, a GFP-FKBP protein is used which does not attract clathrin to the plasma membrane.

Following the dimerization of hook and anchor in response to rapamycin, there is rapid recruitment of clathrin and new clathrin-coated vesicles (CCVs) are formed. These de novo vesicles are recognizable by their intense GFP fluorescence. Their origin from the plasma membrane can also be confirmed by fluorescent antibody labeling of the anchor in live cells. We present here detailed protocols to trigger endocytosis using our hot-wiring method as well as a guide for the imaging and analysis of the vesicles generated.

2 Materials

2.1 Cell Preparation

1. Complete growth medium: DMEM supplemented with 10% FBS, 1% penicillin/streptomycin.
2. Serum-free DMEM.
3. Sterile phosphate buffered saline (PBS).
4. 0.25% trypsin–EDTA.
5. Genejuice transfection reagent (Novagen).
6. 6-well tissue culture plates.
7. Fluorodish imaging dishes (WPI).
8. Plasmid DNA—one plasma membrane "anchor," e.g., CD8-mCherry-FRB (Addgene #100738) or CD8-dCherry-FRB (Addgene #100739) and one clathrin "hook," e.g., FKBP-β2-GFP (Addgene #100726).

2.2 Immuno-fluorescence

1. #1.5 glass coverslips.
2. Fine forceps for coverslip handling.
3. Anti-CD8 primary antibody with or without fluorophore conjugation (Bio-Rad).

4. Green, red, far-red fluorescently labeled secondary antibodies (Life technologies).

5. Unconjugated goat anti-mouse IgG (ThermoFisher).

6. Fixation buffer: 3% paraformaldehyde, 4% sucrose in PBS.

7. Permeabilization solution: 0.1% Triton X-100 in PBS.

8. Glass microscope slides.

9. Mowiol + DAPI mounting media: 2.4 g Mowiol 4–88, 6 g glycerol, 6 ml H_2O, 12 ml 0.2 M Tris-Cl (pH 8.5), 2.5% DABCO, 10 μg/ml DAPI.

2.3 Imaging Equipment

1. Confocal microscope: Nikon Eclipse-Ti with PerkinElmer Ultraview spinning disk. Fitted with 100x oil objective, 405/488/561/640 nm lasers, 2 × Hamamatsu Orca cameras and environmental chamber set at 37 °C. Operated by Volocity software.

2. Immersion oil.

3. Leibovitz's L-15 CO_2-independent medium (no phenol red) supplemented with 10% FBS.

4. Rapamycin (Alfa Aesar) 200 μM in ethanol stock.

5. Ethanol.

6. Pipette for 100 μl.

7. Image analysis and graphing software: Fiji, Igor Pro version 7 (WaveMetrics).

3 Methods

Hot-wired endocytosis is best observed by live-cell imaging. Here, the vesicles appear as bright green spots of FKBP-β2-GFP following rapamycin addition. These vesicles can be isolated using image segmentation methods. An advantage of this technique is that it allows the dynamics of the process to be observed and there is data from the same cell before and after hot-wiring. Another way to image this process is through antibody feeding, and then visualizing the uptake in live or fixed cells. This has the advantage that the antibody-labeled anchor acts as confirmation that the bright green spots are recently derived from the plasma membrane. Ideally this is done in live cells to keep the benefits of that technique, but the number of fluorophores can be limiting. With additional immuno-fluorescence steps, fixed cells can be used to determine the end point of endocytosis triggered by hot-wiring. We describe each of these methods below.

3.1 Cell Preparation

We have used HeLa cells for hot-wiring endocytosis due to their ease of transfection. However, we have also used RPE1 cells and primary cultures of hippocampal neurons, we expect this protocol to be applicable to a range of cell lines. All steps are carried out in a tissue culture hood under sterile conditions.

1. *Day 1*—Seed HeLa cells in 6-well plates, 180,000 cells/well in complete growth medium, 2 ml per well.

2. *Day 2*—Transfect the cells using an appropriate method; chemical transfection with GeneJuice was used here. Mix 100 μl serum-free DMEM with 3 μl GeneJuice (per well for transfection) and incubate at room temperature for 5 min.

3. In a separate 1.5 ml tube, mix 500 ng of each plasmid DNA per well (*see* **Note 1**). Typically these are the anchor and hook constructs, CD8-mCherry-FRB and FKBP-β2-GFP (*see* **Note 2**). If a third plasmid is needed, a further 500 ng of DNA is added, with a proportionate increase in the amount of GeneJuice to 4.5 μl per well in **step 2**.

4. Combine the diluted GeneJuice and DNA, mix gently and incubate at room temperature for 15 min.

5. Replace the growth medium on the HeLa cells with fresh complete growth medium (3 ml for one well of a 6-well plate). At this stage, the cells should be between 60 and 80% confluent. Add 100 μl transfection mixture per well and mix by gentle rocking.

6. Cells are incubated with the GeneJuice-DNA mixture overnight or for a minimum of 5 h, replacing the media with 2 ml complete growth medium at the end of the incubation period.

7. *Day 3*—The cells should be split 1:3 (*see* **Note 3**) onto ethanol-sterilized coverslips in a 6 or 12 well plate, or glass-bottomed imaging dishes such as FluoroDishes.

8. *Day 4*—Cells are ready for imaging or immunofluorescence.

3.2 Live Cell Imaging of Hot-Wired Endocytosis

All imaging was performed using a 100× objective on a spinning disk confocal microscope equipped with 405 nm, 488 nm, 561 nm, and 640 nm lasers. The stage was enclosed in an environmental chamber set to 37 °C. Live cell imaging was performed using a dual camera system for simultaneous observation of GFP and mCherry excited with 488 nm and 561 nm laser lines, respectively.

1. Prepare a ready-to-use stock of rapamycin by adding 16 μl of 200 μM rapamycin (in ethanol) to 1 ml CO_2-independent medium (L-15) and keep at 37 °C. For the control, add 16 μl of 100% ethanol to 1 ml L-15.

2. Replace DMEM for 1.5 ml L-15 + 10% FBS in FluoroDishes containing HeLa transfected with CD8-mCherry-FRB and FKBP-β2-GFP (see **Note 4**).

3. Transfer the dish to the microscope and locate transfected cells using 488 nm (green) and 561 nm (red) lasers. Select the target cell(s), avoiding any with an atypical expression pattern (see **Note 5**).

4. Acquire dual color images at 5 s intervals of the chosen cell for 1 min to record a baseline for the two proteins (see **Note 6**).

5. Without pausing the acquisition, gently add 100 μl of the diluted rapamycin or ethanol solution using a pipette to a final concentration of 200 nM. This step triggers endocytosis.

6. Continue with image acquisition of both channels at the same frame rate until the complete response has been recorded. Successful initiation of hot-wired endocytosis results in the formation of characteristic bright GFP puncta, as shown in Fig. 1.

7. Rapamycin-induced heterodimerization is irreversible within experimental timescales so the dish is not reusable [8].

3.3 Live Cell Imaging of Antibody Labeled Hot-Wired Vesicles

1. Incubate cells expressing CD8-mCherry-FRB and FKBP-β2-GFP with 1:100 anti-CD8 conjugated to Alexa 647 (MCA1226A647, Bio-Rad) or similar far-red label in full DMEM for 30 min at 37 °C.

2. Gently wash the dish with L-15 to remove excess antibody and replace with 1.5 ml L-15 + 10% FBS.

3. Perform live cell imaging as above (see Subheading 3.2, **step 3** onward), with capture of additional far red (640 nm excitation) images at each time point.

3.4 Imaging Antibody Labeled Hot-Wired Vesicles in Fixed Cells

1. To HeLa cells on coverslips expressing CD8-dCherry-FRB and FKBP-β2-GFP, add 1:1000 unlabeled anti-CD8 (MCA1226, Bio-Rad) in complete growth medium at 37 °C, incubate for 30 min.

2. Trigger endocytosis by adding rapamycin to a final concentration of 200 nM and return to 37 °C. For a maximum response, incubate with rapamycin for 20 min.

3. Place cells on ice to prevent further endocytosis and wash with ice-cold complete growth medium.

4. Add Alexa 647-conjugated secondary antibody in complete growth medium for 1 h at 4 °C to label the CD8 antibody remaining at the cell surface (see **Note 7**).

5. Wash cells 3 times with ice-cold complete growth medium.

6. Optional—Add unconjugated anti-mouse IgG at 1:100 in DMEM for 30 min to saturate any remaining antibody binding

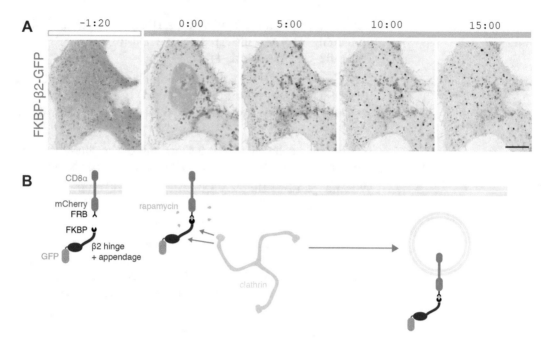

Fig. 1 Live cell imaging of hot-wired CME. **(a)** Time-lapse images showing the rerouting of FKBP-β2-GFP to CD8-mCherry-FRB at the plasma membrane following rapamycin addition at $t = 0$ min (filled bar). The number of vesicles increases over time, peaking at around 10 min. Time shown as min:sec, scale bar = 10 μm. **(b)** Schematic diagram showing the arrangement of β2 hook and CD8 anchor during the hot-wiring process. Addition of rapamycin induces dimerization of the two proteins causing the recruitment of clathrin and the formation of new clathrin-coated vesicles. Figure adapted from [4]

sites at the plasma membrane and prevent their detection in **step 9**.

7. Fix the cells using 3% PFA, 4% sucrose in PBS for 15 min.

8. Permeabilize the cells using 0.1% Triton X-100 in PBS for 10 min. Cells may now be brought to room temperature.

9. Then, add Alexa 568-conjugated secondary antibody in complete growth medium for 1 h to label any anti-CD8 that was internalized during the experiment.

10. After washing with PBS and rinsing in dH₂O, the coverslips can be mounted using Mowiol + DAPI and allowed to dry overnight before storage at 4 °C.

11. Image the slides using GFP to determine which cells were transfected and formed hot-wired vesicles. It is preferable to use 100× and take z-stacks of the entire cell volume at ~0.5 μm intervals to accurately localize the proteins.

12. The cell surface CD8 population is stained far-red, internalized CD8 originating from the plasma membrane is stained red and will colocalize with FKBP-β2-GFP. An example is shown in Fig. 2.

FKBP-β2-GFP   Internal CD8   Surface CD8

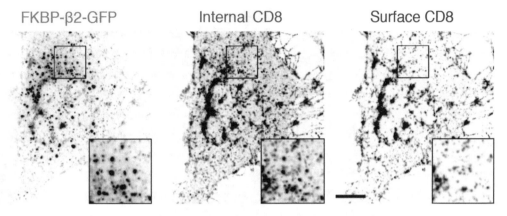

Fig. 2 Protein localization in fixed cells following antibody feeding. Confocal images of antibody uptake in response to hot-wired CME. Where there is colocalization between FKBP-β2-GFP and Internal anti-CD8, this represents an endocytic event that has originated from the plasma membrane between rapamycin addition and fixation. Where there is internal and surface anti-CD8 colocalization, this identifies the membrane CD8 population rather than internalized vesicles. Scale bar = 10 μm, inset shows 2× zoom. Figure adapted from [4]

3.5 Analysis of Live Cell Images of Hot-Wired Endocytosis

The bright green puncta of FKBP-β2-GFP which form after endocytosis is triggered can be isolated using segmentation methods. Analysis is semiautomated and performed in Fiji (Image J) and Igor Pro. Custom-written code for automated processing manually obtained Fiji outputs using Igor Pro is available here: https://github.com/quantixed/PaperCode/tree/master/Wood2017.

1. Open/Import the image stack in Fiji, the "hook" channel (FKBP-β2-GFP) is analyzed.

2. First, if significant bleaching has occurred during the movie, correct it using the simple ratio method. This improves the accuracy of the subsequent thresholding.

3. Take a small region of interest (ROI) in the cytoplasm, avoiding any obvious bright structures and the nucleus in all frames of the movie. Measure the mean pixel intensity within the ROI in each frame using the multimeasure function; the value should sharply decrease at the point of rerouting when the GFP signal moves from cytoplasm to the plasma membrane (*see* **Note 8**).

4. Draw an ROI around the perimeter of the cell and save in ROI manager.

5. Threshold the entire image to isolate only the bright green spots that form following rapamycin addition (*see* **Note 9**). This should exclude any puncta visible in the pre-rapamycin frames and also the increase in green fluorescence at the membrane following rapamycin. This will result in a binary image with the vesicles and background marked by pixels with a value of 255 and zero respectively (*see* **Note 10** and Fig. 3).

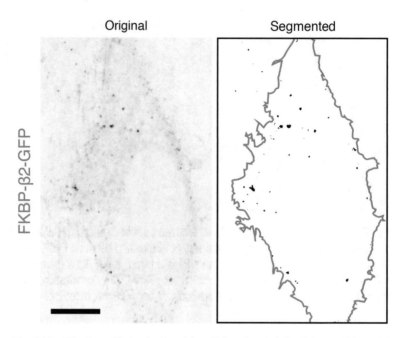

Fig. 3 Identification of hot-wired vesicles. Using thresholding it is possible to differentiate hot-wired vesicles by segmenting spots that are above a user-defined intensity. These thresholded images can then be quantified to determine the extent of vesicle formation. Figure adapted from [4]

6. Restore the whole cell ROI (from **step 4**) to this mask and use the multimeasure tool to produce data on every frame of the stack. The value used for analysis is the raw integrated density (RawIntDen) which gives the sum of the pixel values within the ROI. As all objects identified in the threshold have the same pixel value, this is equivalent to the number of pixels in these objects multiplied by 255. From the same data set, also record the area of the ROI.

7. Import into IgorPro three sets of values per cell, mean green intensity ("g wave"), RawIntDen (puncta wave or "p wave"), and area wave (constant, "a wave"). If these are stored in Excel, use a worksheet for each condition to be analyzed and use a logical naming system for each column, e.g., WT_1_g, WT_2_g, WT_1_p, WT_2_p, WT_1_a, WT_2_a. Igor will import all the data from the workbook and then present a GUI to help the user determine for every cell the time at which the green intensity has fallen by 50% (Fig. 4a). This provides the time point within the movie where the cell is halfway through the rerouting process (*see* **Note 11**). This timepoint is used for offsetting the data for averaging.

8. The code then plots the p wave vs time for each cell in a line graph. For functional clathrin hooks there is an increase following rapamycin addition, whereas for nonfunctional hooks

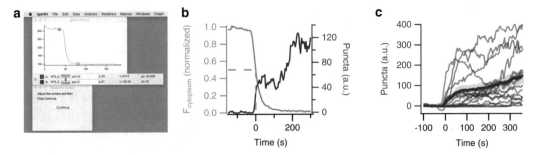

Fig. 4 Results of live-cell image analysis. (**a**) View of GUI to find the 50% rerouting point for time-alignment. (**b**) Increase in puncta over time ("p wave," right axis, dark blue), normalized to pre-rapamycin addition. The corresponding decrease in cytoplasmic GFP fluorescence ("g wave," left axis, light gray) used to time-align the puncta wave (dashed lines). Same cell as in **a**. (**c**) Multiple time-aligned, normalized puncta traces from different cells. Average ± s.e.m. is shown as a thick line with shading

and control proteins, the line stays flat. For each p wave, $t = 0$ is redefined as the time point where the cytoplasmic green fluorescence has fallen to 50%, as determined in **step 7**. This normalizes the response times from all cells to allow them to be averaged.

9. Prior to averaging, each p wave should be normalized to cell size by dividing by the area of the ROI containing the vesicles, as drawn in **step 4**.

10. The final normalization step is to adjust the y-offset of the p wave to compensate for any cells where significant background fluorescence has been picked up by the threshold. This is done by taking an average RawIntDen value from the time points before $t = 0$ and then subtracting this from all values in that p wave. Example from a single cell of normalized p and g waves are shown in Fig. 4b.

11. Having corrected for timing, area, and intensity differences, these final p wave values can be averaged to determine the mean change in RawIntDen for that group of cells (Fig. 4c).

4   Notes

1. We have a preference in our lab for Genejuice, which we find gives the highest transfection efficiency (other transfection methods can be used). When there was a large difference in the transfection efficiency of the constructs, it was beneficial to add disproportionately more of the weakly expressing protein, e.g., 250 ng + 750 ng.

2. We observed better localization of the anchor to the plasma membrane when CD8 and FRB domains were separated by mCherry. In instances where a nonfluorescent anchor was

required, a mutant of mCherry (K70N termed dark Cherry or dCherry) was used instead [9]. All alternative anchor proteins we tested worked well. For the clathrin hook we use a fragment of β2 adaptin (hinge and appendage) as this is ordinarily located at the plasma membrane as a subunit of AP2 and is known to strongly bind clathrin [10]. Again, this is exchangeable for other clathrin-binding proteins, although the ability to bind clathrin in the test-tube does not necessarily mean that the protein will work as a clathrin hook. The precise rules for how hooks engage clathrin in a functional context are currently unclear.

3. Splitting confluent transfected cells 1 well into 3 wells worked well but can be increased or decreased depending on cell confluence and transfection efficiency.

4. We prefer to image cells in phenol red-free CO_2-independent media without regulating CO_2 levels in the microscope incubation chamber. Media without phenol red reduces autofluorescence.

5. We found that allowing the dish to equilibrate for 5–10 min in the environmental chamber greatly reduced focus drift. It is best to select cells for imaging on the basis of their mCherry fluorescence. It is important to have good localization of CD8 to the plasma membrane and the minimum amount of intracellular fluorescence. Intracellular deposits of CD8-mCherry-FRB were always present although these do not prevent rerouting to the plasma membrane. Cells without a defined plasma membrane or with visible endoplasmic reticulum fluorescence should be avoided. Furthermore, rerouting efficiency of GFP signal was reduced in very dim and very bright cells, so these should also be avoided.

6. Typically, 12 images at 5 s per frame were taken prior to rapamycin addition and then a further 200 frames were taken to capture the entire response. The initial rerouting response occurs in <5 s and vesicle production peaks within the first 5 min, therefore if dynamics are required it is not recommended to acquire images slower than 10 s per frame. If vesicle tracking is required, a minimum of 2 s per frame is recommended. The illumination intensity should be kept as low as possible to reduce bleaching within the movie and prevent image saturation due to the brightness of the induced vesicles.

7. Labeling of internal anti-CD8 can also detect remaining surface binding sites, therefore wherever there is colocalization of internal and cell surface CD8 labeling; these areas are excluded from analysis. It is essential to avoid permeabilizing the cells before the surface population has been labeled.

8. The time taken for individual cells to respond to rapamycin is variable and this step is necessary if an average response is to be calculated from several cells, so that they are time-aligned to the moment of rerouting.

9. These bright spots are presumed to be hot-wired vesicles as they do not appear prior to rapamycin addition, when a non-clathrin-binding hook is used, or following clathrin siRNA treatment. They also colocalize with clathrin, anti-CD8 fed extracellularly, and by electron microscopy they represent clathrin-coated vesicles.

10. Thresholding the image can be difficult. It requires images with good dynamic range. It is easiest to isolate the bright green spots at mid-point of the movie because the final frames are often affected by bleaching. Once the minimum and maximum intensities have been determined for the chosen frame, ensure that there are not a significant number of unwanted structures in the earlier, brighter frames and that not too many GFP spots are lost in the later frames. If irregularly shaped bright objects that are too large to be hot-wired vesicles are included, Fiji's "Analyze particles..." function may be used. This will identify features of a user-defined size and circularity that closely matches the vesicles and can create a binary mask to be used in the same way as the thresholding output. Attempts to automate the thresholding task were unsuccessful due to the complexity of the images, and therefore it is essential to maintain the same inclusion criteria between samples and for the analyzer to be blind to the conditions of the images.

11. The 50% value was determined by a GUI where g waves were presented to the user and then the maxima (before rapamycin) and minima (after rapamycin) selected. These values were used to determine the time at which the midpoint was reached

12. Due to the very rapid response (within 1–2 frames), the midpoint sufficiently represents the time point of rerouting.

Acknowledgments

L.A. Wood was supported by the Medical Research Council Doctoral Training Partnership grant (MR/J003964/1).

References

1. Ehrlich M, Boll W, Van Oijen A, Hariharan R, Chandran K, Nibert ML, Kirchhausen T (2004) Endocytosis by random initiation and stabilization of clathrin-coated pits. Cell 118(5):591–605. https://doi.org/10.1016/j.cell.2004.08.017

2. Taylor MJ, Perrais D, Merrifield CJ (2011) A high precision survey of the molecular dynamics of mammalian clathrin-mediated endocytosis. PLoS Biol 9(3):e1000604. https://doi.org/10.1371/journal.pbio.1000604

3. Dannhauser PN, Ungewickell EJ (2012) Reconstitution of clathrin-coated bud and vesicle formation with minimal components. Nat Cell Biol 14(6):634–639. https://doi.org/10.1038/ncb2478

4. Wood LA, Larocque G, Clarke NI, Sarkar S, Royle SJ (2017) New tools for "hot-wiring" clathrin-mediated endocytosis with temporal and spatial precision. J Cell Biol. https://doi.org/10.1083/jcb.201702188

5. Kelly BT, Graham SC, Liska N, Dannhauser PN, Honing S, Ungewickell EJ, Owen DJ (2014) Clathrin adaptors. AP2 controls clathrin polymerization with a membrane-activated switch. Science 345(6195):459–463. https://doi.org/10.1126/science.1254836

6. Rivera VM, Clackson T, Natesan S, Pollock R, Amara JF, Keenan T, Magari SR, Phillips T, Courage NL, Cerasoli F Jr, Holt DA, Gilman M (1996) A humanized system for pharmacologic control of gene expression. Nat Med 2(9):1028–1032

7. Kozik P, Francis RW, Seaman MN, Robinson MS (2010) A screen for endocytic motifs. Traffic 11(6):843–855. https://doi.org/10.1111/j.1600-0854.2010.01056.x

8. Banaszynski LA, Liu CW, Wandless TJ (2005) Characterization of the FKBP.rapamycin. FRB ternary complex. J Am Chem Soc 127(13):4715–4721. https://doi.org/10.1021/ja043277y

9. Subach FV, Malashkevich VN, Zencheck WD, Xiao H, Filonov GS, Almo SC, Verkhusha VV (2009) Photoactivation mechanism of PAmCherry based on crystal structures of the protein in the dark and fluorescent states. Proc Natl Acad Sci U S A 106(50):21097–21102. https://doi.org/10.1073/pnas.0909204106

10. Owen DJ, Vallis Y, Pearse BM, McMahon HT, Evans PR (2000) The structure and function of the beta 2-adaptin appendage domain. EMBO J 19(16):4216–4227. https://doi.org/10.1093/emboj/19.16.4216

Chapter 8

Real-Time Endocytosis Measurements by Membrane Capacitance Recording at Central Nerve Terminals

Xuelin Lou

Abstract

Endocytosis is fundamental to cell function. It can be monitored by capacitance measurements under patch-clamp recordings. Membrane capacitance recording measures the cell membrane surface area and its changes at high temporal-resolution and sensitivity, and it is a powerful biophysical approach in the field of exocytosis and endocytosis. A popular one is the frequency domain method that entails processing passive sinusoidal membrane currents induced by a sinusoidal voltage. This technique requires a phase-sensitive detector or "lock-in amplifier" implemented in hardware or software during patch-clamp recordings. It has been widely used in many secretory cells, but its application directly at central presynaptic terminals is technically challenging. We have applied this technique to study synaptic endocytosis in the calyx of Held, a large glutamatergic synaptic terminal, as well as mouse pancreatic β-cells. The presynaptic capacitance measurements provide a unique alternative to measuring transmitter release and presynaptic endocytosis. Here, we describe this method at the calyx of Held in acute brain slices and provide a practical guide to obtaining high quality capacitance measurements at presynaptic terminals.

Key words Capacitance measurement, Endocytosis, Secretion, Presynaptic terminals, Patch-clamp, Calyx of Held

1 Introduction

Endocytosis is a fundamental process by which a living cell uptakes exterior materials via plasma memebrane invagination and pich-off. It is regulated by an array of molecules and cellular signaling. Membrane capacitance measurements have been widely used in the study of exocytosis and endocytosis in live cells. During vesicle trafficking, vesicles fuse with the plasma membrane and increase the cell surface area; endocytosis retrieves membrane and decreases surface area. This was first demonstrated by Neher and Marty using capacitance measurements in chromaffin cells in 1982 [1]. Membrane capacitance is proportional to certain membrane surface

The original version of this chapter was revised. A correction to this chapter can be found at
https://doi.org/10.1007/978-1-4939-8719-1_19

Laura E. Swan (ed.), *Clathrin-Mediated Endocytosis: Methods and Protocols*, Methods in Molecular Biology, vol. 1847,
https://doi.org/10.1007/978-1-4939-8719-1_8, © Springer Science+Business Media, LLC, part of Springer Nature 2018

area (9 fF/μm^2) in different cell types (including neurons) [2], and it provides a method to monitor membrane area and its changes. This approach has been widely used in many types of secretory cells with a round shape [3], such as mast cells [4]. It has also been applied to neurons that contain ribbon synapses, such as retina bipolar cells [5, 6], photoreceptors [7], and hair cells [8, 9].

Capacitance measurements in central synapses within intact neural circuits are more difficult, largely due to their small sizes and the challenges of brain slice recordings. However, presynaptic capacitance recording provides a unique method for synaptic exocytosis and endocytosis in a native neural circuit in the brain [10]. Moreover, it offers a direct readout for transmitter release and plasticity without relying on postsynaptic factors. This is an advantage over traditional methods of monitoring postsynaptic currents/ potentials because the latter is an indirect readout of transmitter release and may be affected by postsynaptic receptor desensitization and saturation [11, 12]. In addition, a variety of small molecules of interest, such as Ca^{2+} indicators, caged molecules [13], small signaling molecules and peptides [14] can be dialyzed into presynaptic terminals to study their function. Therefore, despite the technical challenges, these advantages make this approach an important tool for synaptic transmission. The method has been applied to many different types of presynaptic terminals in the brain, such as the calyx of Held [10, 12], the end-bulb of Held [15], pituitary nerve terminals [16], and mossy fiber terminals [17]. Many of these terminals have more complex structures (i.e., long axon) equivalent to multiple compartments.

Both time domain and frequency domain methods have been developed for membrane capacitance measurements [18, 19]. The time domain method is simple; it determines cell membrane parameters using amplitude and time course of membrane current relaxations induced by square voltage waves. Large changes of access resistance during recordings are tolerated by this approach. The frequency domain method utilizes the phase shift between an input sinusoidal voltage wave and the resulting sinusoidal current output to calculate the complex impedance of the electric circuit using a lock-in amplifier (implemented either in hardware or software). This approach is largely simplified by digital instruments with a fully computer-controlled patch-clamp amplifier and a software-based phase sensitive detector in "LockIn" extension software [20]. The virtual instruments provide a full control of experimental operations and great flexibility in adjusting parameters. Two techniques can be applied using a lock-in amplifier: the piecewise-linear (PL) technique and Sine + DC (SDC) technique [18]. In both modes, voltage-dependent conductance must be avoided so that only passive capacitive currents remain during capacitance measurements. The SDC technique has some advantages over the PL technique since it generates all three parameters including membrane capacitance (Cm), membrane conductance

(Gm), and resistance conductance (Gs), rather than their relative changes in the PL technique. Thus, it has become a popular approach for capacitance measurement.

We have successfully applied capacitance measurements to study endocytosis in presynaptic terminals in brain slices [10], in addition to pancreatic β-cells [21]. The calyx of Held is a glutamatergic presynaptic terminal, and it contains conventional active zones and innervates the principal neurons of the medial nucleus of the trapezoid body (MNTB) in the auditory brainstem. It is accessible to direct patch-clamp [22, 23] and becomes a powerful model synapse to study central synapses [24, 25]. In the past decade, we have used this central nerve terminal to study presynaptic Ca^{2+} sensing and vesicle fusion [13, 26, 27], plasticity [28–30], endocytosis [10, 31, 32], and synaptic development [33]. In this chapter, we describe the step-by-step method for presynaptic capacitance measurements in order to study synaptic vesicle exocytosis and endocytosis at the calyx of Held in acute brain slices.

2 Materials

2.1 Glass Pipette

1. A patch-clamp pipette puller (PIP-6, HEKA, Germany).

2. Thick-wall borosilicate glass capillaries, ID = 1.4 ± 0.05 mm, OD = 2:00 ± 0.05 mm, length = 80 ± 3 mm, roundness = 0.05.

3. Sylgard 184 silicone elastomer kit (WPI instrument, Sarasota, FL). Sylgard coating material is prepared by mixing silicone elastomer base with curing agent at 10:1 ratio (*see* **Note 1**).

4. A microforge station for pipette coating and heat-polishing (CMP-2, ALA, Farmingdale, NY). It is integrated with an inverted microscope with 20× and 40× objectives, and a small hot-air jet that is heated by a coil is placed adjacent to the pipette tip (*see* **Note 2**).

5. Heated air blower.

2.2 Electrophysiology Setup for Capacitance Measurements

Capacitance recordings are performed under whole-cell patch-clamp, which requires standard components for patch-clamp recordings. We use our current electrophysiology system as an example.

1. A patch-clamp amplifier (EPC-10, HEKA, Germany) (*see* **Note 3**).

2. A computer with Patchmaster and LockIn extension for capacitance measurements.

3. A vibration isolation table with a Faraday cage.

4. An inverted fluorescence microscope with 5× and 60× (NA = 0.95, long working distance, water immersion) objectives and IR DIC illumination.

5. A CCD camera and a monitor for patch-clamp experiments.

6. An X-Y translator for the microscope.

7. Micromanipulators (MP-285, Sutter).

8. AgCl electrode.

9. A recording chamber for brain slices (Warner Instruments) and multiple-line slice perfusion system (ALA).

10. Standard extracellular solution (ES): 120 mM NaCl, 2.5 mM KCl, 25 mM NaHCO$_3$, 1.25 mM NaH$_2$PO$_4$, 2 mM CaCl$_2$, 1 mM MgCl$_2$, 25 mM glucose, 3 mM myo-inositol, 2 mM Na-Pyruvate, 0.4 mM (L-) ascorbic acid (pH 7.4, when continuously bubbled with 95% O$_2$ and 5% CO$_2$). The ES is freshly prepared daily with stock solution (*see* **Note 4**).

11. Intracellular/pipette solution (IS): 137 mM Cs-gluconate, 10 mM Hepes, 20 mM TEA-Cl, 5 mM Na$_2$-phosphocreatinine, 4 mM Mg-ATP, 0.3 mM Na$_2$GTP, 0.2 mM Cs-EGTA (pH = 7.2) (*see* **Note 5**).

12. 10 mM TEA-Cl (Tetraethylammonium chloride), 1 µM tetrodotoxin in ES.

2.3 Brain Slices

1. A vibratome (VT-1200, Leica) for acute brain slices (*see* **Note 6**).

2. A water bath maintained at 37 °C.

3. A slice storage chamber.

4. A gas tank containing 95% O$_2$ and 5% CO$_2$.

5. Tissue dissecting tools.

6. Tissue dissecting solution: 120 mM NaCl, 2.5 mM KCl, 25 mM NaHCO$_3$, 1.25 mM NaH$_2$PO$_4$, 0.1 mM CaCl$_2$, 3 mM MgCl$_2$, 25 mM glucose, 3 mM myo-inositol, 2 mM Na-Pyruvate, 0.4 mM (L-)ascorbic acid.

7. Slice recovery solution (the same as the standard ES).

3 Methods

3.1 Patch Pipette Preparation

1. Patch pipettes are freshly prepared before each experiment from borosilicate capillaries. They are fabricated using a pipette puller with appropriate pulling parameters to produce pipettes with about 3–5 MΩ resistance when filled with IS.

2. Coat each pipette with premixed Sylgard 184. Apply Sylgard to pipette cone so that it covers pipette cone as close as possible to but not on the pipette tip (*see* **Note 7**). This produces an even layer of insulating coat on the pipette that directly contact with bath solution during recordings, which is critical for reducing capacitance recording noise.

3. Dry the Sylgard layer immediately after coating with a heated air blower (*see* **Note 8**).

4. Fire polish pipette tip on the polishing station, monitor it under a microscope (*see* **Note 9**). With experience, pipette resistance can be roughly estimated according to the shape and diameter of pipette tip under a microscope.

5. Keep coated pipettes in a dust-free chamber (with a lid) for experiments on the same day (*see* **Note 10**).

3.2 Acute Brain Slice Preparation

1. Prepare dissecting tools, 37 °C water bath, cold dissecting buffer, and a mouse pup. We normally use pups at their postnatal day 8–10 (P8–10).

2. Decapitate and remove the brain. Rapidly dissect out brainstem with caution, do not to damage MNTB regions (*see* **Note 11**); rapidly immerse tissue into ice-cold dissecting buffer presaturated with 95% O_2 and 5% CO_2, trim and rinse the tissue block.

3. Mount the brainstem block in a cold slicing chamber, add ice-cold tissue dissecting solution, adjust orientation and height of mounting base, and move the blade slightly above the MNTB level.

4. Cut coronal slices at 180–200 μm thickness with a speed of 0.1 mm/s and vibration amplitude of 1 mm (*see* **Note 6**). Slices containing MNTBs are collected. Normally, four good slices can be obtained from each mouse at this age.

5. Transfer slices to a storage chamber containing prewarmed (37 °C) and continuously bubbled ES with 95% O_2 and 5% CO_2 for ~30 min. Then, the slice chamber is moved out of the water bath and maintained at room temperature with continuous bubbling.

3.3 Patch-Clamp at Presynaptic Terminals

1. Thaw an aliquot of frozen intracellular solution and place it on ice.

2. Place a slice in the recording chamber and put a slice holder on top of it, then load the chamber on the microscope stage. Continuously perfuse the slice with ES containing 1 μM tetrodotoxin and 10 mM TEA-Cl under continuous 95% O_2 and 5% CO_2 bubbling.

3. Visually screen MNTB regions in the brain slice and identify a presynaptic terminal (Fig. 1) suitable for patch-clamp. This is performed using the monitor connected the microscope camera (infrared DIC illumination, 60× objective) (*see* **Note 12**).

4. Fill a patch pipette with 20 μL intracellular solution that contacts with the AgCl electrode in the pipette holder (*see* **Note 13**). Locate pipette to nerve terminals, obtain a gigaohm seal, and compensate fast capacitance.

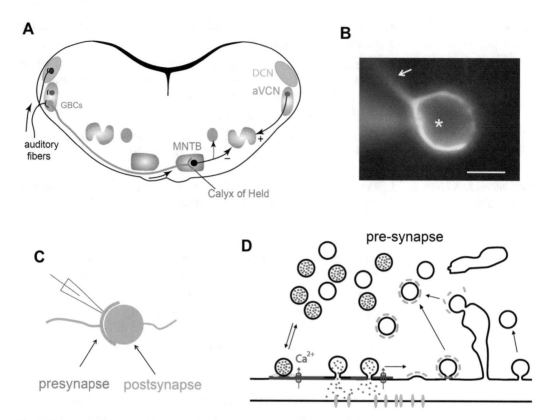

Fig. 1 The calyx of Held, a central synapse suitable for capacitance measurements. (**a**) The location of MNTB and the calyx of Held in the auditory brainstem slice. (**b**) A live calyx of Held illuminated by the fluorescence of Fura-2 infused through a patch pipette (which is above the focus plan on the left). Note the thin layer of pre-synaptic membrane wrapped on a principal MNTB neuron (*). Arrow indicates the connecting axon, which travels out of the focus plan. (**c, d**) Diagrams of patch-clamp recording at the calyx of Held (**c**) and vesicle cycle at an active zone (**d**). Multiple endocytosis pathways are illuminated and empty circles represent newly formed vesicles through endocytosis. Scale bar in (**b**): 10 μm

5. Lower the bath solution level as low as possible by adjusting the level of suction pipette tip in the recording chamber, and maintain a stable fluid level throughout the recording (*see* **Note 14**).

6. Set holding membrane potential to −80 mV, recompensate fast capacitance and rupture patch membrane to achieve whole-cell patch-clamp mode, compensate slow capacitance and access resistance before starting recording. Access resistance should remain low and stable (*see* **Note 15**).

3.4 Capacitance Measurements of Endocytosis at Real-Time

1. Activate the lockIn extension of Patchmaster according to the manufacture's manual and select the "Sine + DC" mode.

2. Select a stimulation protocol for a specific experimental purpose in a pulse generator that is prepared and tested in advances (*see* **Note 16**).

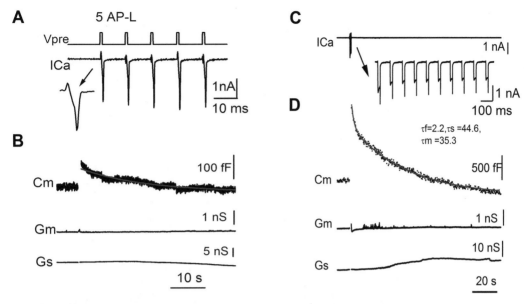

Fig. 2 Synaptic exocytosis and endocytosis revealed by capacitance measurements at the calyx of Held from mouse brain slices. (**a**) Presynaptic stimulation and Ca^{2+} currents induced by a train of action potential like (AP-L) pulses. (**b**) Synaptic exocytosis and endocytosis recorded by capacitance measurements. Experiments were performed with the frequency domain ("sine+ DC") technique under whole-cell patch-clamp (holding potential = −80 mV); five AP-L pulses (1 ms, 20 mV at 10 ms interval, top in **a**) were applied to a presynaptic terminal. Note that the small Cm jump and its subsequent decay to baseline (τ = 14 s), without correlated changes in Gm and Gs. (**c, d**) Synaptic exocytosis and endocytosis following intense stimulation (10 pulses at 10 Hz, each pulse was 10 mV and 20 ms). Note the large presynaptic Ca^{2+} influx (**c**), Cm jump, and rapid endocytosis (**d**), in which endocytosis is better fitted with a double exponential (τ_{fast} = 2.2 s, τ_{slow} = 44.6 s). Gm and Gs showed no correlated changes as Cm. Data were adapted from [10], and the first 400 ms Cm following the stimulation was not blanked here since it contain signal irrelevant to transmitter release

3. Apply the stimulation to the presynaptic terminal and record membrane capacitance and their changes with time. "Sine + DC" capacitance mode in "LockIn" extension produces multiple parameters including Cm, Gm and Gs. They can be acquired and displayed on the computer screen at high temporal resolution (Fig. 2) (*see* **Note 17**).

4. For long-term recordings of membrane capacitance, a low time resolution recording is very useful, and this is often sufficient for a slow form of endocytosis, e.g., clathrin-mediated endocytosis. Low time resolution capacitance recordings use continuously repetitive short segments of voltage sine waves superimposed on membrane potential before and after stimulation. Each segment gives rise to an averaged Cm, Gs, and Gm that can be monitored in real-time during recordings (Fig. 3). High resolution and low resolution capacitance recordings are often combined for different experimental purposes (Fig. 3).

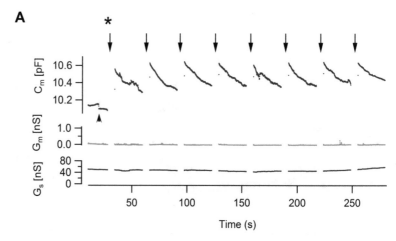

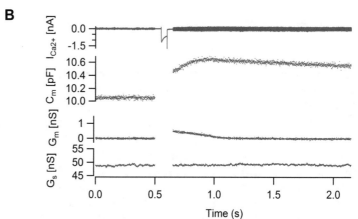

Fig. 3 Capacitance measurements at presynaptic terminals with the combination of low- and high-temporal resolution recordings. (**a**) A low time resolution Cm recording at a presynaptic terminal. The calyx of Held is stimulated repeatedly with a depolarizing pulse (0 mV, 50 ms) (arrows). Cm, together with Gs and Gm, was monitored at 63 ms interval with a 50 ms segment of sine waves before and after each pulse. Note the robust, repetitive exocytosis (Cm jump) and subsequent endocytosis (Cm decay) following each stimulation, without parallel changes in Gm and Gs. Arrow head indicates a rapid downward step in Cm but not in Gm and Gs. (**b**) High resolution Cm measurements of the first stimulation (*) recorded in (**a**)

3.5 Data Analysis of Capacitance Data for Endocytosis

1. Capacitance measurements can be compromised by other factors, and their impacts depend on cell types, stimulation strength, and experimental conditions. Capacitance data must be analyzed with caution to avoid misinterpretations.

2. During capacitance measurements with the frequency domain technique, some artefacts may contribute to apparent capacitance changes that are irrelevant to or mask true surface area changes from exocytosis or endocytosis. One of them is voltage-dependent activation of ion channels that causes frequency-

dependent and phase-shifted sinusoidal currents [34]; capacitance measurements are unreliable when membrane conductance is activated, such as during depolarization in excitatory cells. Another common source of artefacts is gating currents from charge redistribution in ion channel proteins during conformational transitions. For example, Na$^+$ channel gating currents from inactivation in chromaffin cells causes a transient voltage-dependent apparent capacitance change (τ = 606 ms), which can be blocked by 0.2 mM dibucaine, a drug that blocks Na$^+$ channels and their gating currents [35, 36].

3. To evaluate the quality of capacitance data, capacitance (Cm) and conductance traces (Gm and Gs) should always be monitored together.

4. At the calyx of Held, a transient capacitance change directly recorded immediately following a depolarizing pulse contains signals that is insensitive to SNARE cleavage and thus irrelevant to presynaptic transmitter release [10], which is omitted from analysis in general (*see* **Note 18**).

5. Small linear capacitance drifts, with an unknown source, can be corrected offline by linear subtraction of a baseline generated from a linear regression of sufficient length of capacitance trace before stimulation (*see* **Note 19**).

4   Notes

1. The viscosity of a Sylgard 184 mixture depends on the ratio of two components. A 10:1 ratio leads to an even, thin layer of smooth coating on the pipette tip. If a thicker layer is required, the ratio should be decreased (for example 7:1) to increase the viscosity of the mixture. In addition to repeating the coating and dry cycles. The mixed Sylgard can be stored at −20 °C in small aliquots (500 μL) for several months.

2. If wax is used for coating, an air-heating blower is not necessary since liquid wax at high temperature will dry in a few seconds at room temperature. This property often leads to uneven surface coating of pipettes.

3. We are currently using an EPC10–2, a double patch-clamp amplifier (HEKA instrument, Germany). All the EPC9 amplifiers or later versions are fully controlled by computers through the dedicated software (e.g., PULSE or Patchmaster) and its "LockIn" extension implemented in the software package. This patch-clamp system provides convenient virtual instruments for capacitance measurements, with both PL and SDC approaches available. If other versions or models of amplifiers (such as EPC7, EPC8, Axonpatch200B, multipleClamp700B,

or AxoClamp900A) are used, an extra lock-in amplifier (for example, SR830, CA) is needed.

4. Two different types of stock solution are prepared. $10\times$ stock solution (#1, stored at 4 °C) contains NaCl, KCl, NaHCO$_3$, NaH$_2$PO$_4$. $200\times$ additional solution (#2, stored at −20 °C at 5 mL per unit) contains myo-inositol, Na-Pyruvate, ascorbic acid. To prepare 1 L standard ES, first dilute #1 solution 100 mL with water under continuous 95% O$_2$ 5% CO$_2$ bubbling, add glucose powder and 10 mL predissolved #2 stock under bubbling, and make the final volume to 1 L.

5. This solution is for simultaneous recordings of capacitance and presynaptic Ca^{2+} current. If Ca^{2+} currents are not required, standard IS can be used: 137 mM K-gluconate, 10 mM Hepes, 5 mM Na$_2$-phosphocreatinine, 4 mM MgATP, 0.3 mM Na$_2$GTP, 0.1 mM K-EGTA (pH = 7.2).

6. Vibratome with minimal vibration along the z-direction is critical to produce high quality tissue slices. This decreases neuron damage and preserves intact morphology of presynaptic terminals which is important for successful presynaptic capacitance recordings. The parameters for cutting slices should be optimized according to brain region, slice thickness, animal age and species.

7. Make sure that Sylgard does not contact pipette tip, since Sylgard prevents formation of gigaohm seals during patch clamping. Sylgard coating is critical to reduce recording capacitance noise. A thicker layer of Sylgard can be considered in some cases when noise is high, and it can be achieved by multiple repeated coating–drying cycles or using a Sylgard mixture with higher viscosity (*see* also **Note 1**).

8. A wet Sylgard mixture may creep to and block the pipette tip if it is left too long without drying.

9. This step is optional if gigaohm seals are not an issue during patch clamp. Fire polishing makes the edge of pipette tip smoother and cleaner, and thus promotes gigaohm seals and decrease recording noise. The heating filament tip should be precoated with glass to avoid metal evaporation onto the pipette tip during polishing.

10. Storing coated pipettes for 2 days or longer is not recommended, since this often cause a problem for gigaohm seals. Pipettes should be prepared on the same day of experiments.

11. This step should be done as quickly as possible; avoid long dissecting times without reduced tissue temperature. MNTBs are located right under the surface of ventral side of brainstem, and are prone to damage during the dissection.

12. The identification of calyx of Held in brain slices is relatively easy in mice before hearing onset (i.e., postnatal day 7–10). In many cases, a thin layer of presynaptic membrane covering a large portion of a MNTB neuron is evident under a good DIC microscope. Nerve terminals located just beneath slice surface often give good results. In most calyces, a double-exponential function is necessary to accurately describe the relaxation of passive membrane current in response to a 10 mV pulse (from −80 mV to −70 mV), and the fast and slow decay time constants in P8–10 rats are 0.153 ms and 0.73 ms, respectively [37]. However, the capacitance results are not significantly different in experimental data between calyces with short and long axons, and the impact of a second slow charging membrane compartment is very limited in most cases [37, 38]. Capacitance measurements are also validated in other complex terminals that are approximately equivalent to a circuit of spherical cell coupled with a cylinder process, such as pituitary terminals [16] and mossy fiber terminals [17].

13. This reduces the recording noise of capacitance measurements.

14. Minimize the length of the pipette tip in contact with bathing fluid during capacitance recordings. This is critical to decrease capacitance noise and avoid large capacitance baseline drift and fluctuations. Maintaining a stable, low level of bathing fluid and pipette coating with a smooth and thick layer of Sylgard are two effective ways to obtain low noise capacitance recordings.

15. Maintaining a constant access resistance helps stabilize endocytosis recordings. Endocytosis studies often require continuous capacitance recordings for minutes or even longer without interruptions, and this makes endocytosis more challenging than exocytosis which normally is complete within a second. A small access resistance (12 MΩ or less) under whole-cell often remains stable for a longer time and helps reduce capacitance noise. 3–4 MΩ pipettes often produce good results in our experiments, without a significant impact on gigaohm seals.

16. It is a good idea to prepare stimulation protocols before the experiment. The major difference between normal patch-clamp recordings and capacitance measurements is applying a sine wave voltage superimposed on membrane holding potential and computing capacitance information from the resulting sinusoidal currents and other parameters of the circuit at a give time point. A fully computer-controlled patch-clamp system makes these processes much easier than before. To configure stimulation, use "StimScale, LockIn" in "Stimulus- > DA" panel, choose current signal (Imon-1) in AD panel and LockIn Cm/Gm/Gs. Then, apply a sine wave with optimal parameters on a hyperpolarized membrane holding potential (−80 mV). Peak amplitude and points per cycle of sine waves

are no less than 10 in general, to decrease the noise and to insure capacitance accuracy. We often set the frequency of sine waves as 800–1000 Hz, points/cycle as 10–30, peak amplitude as 10–30 mV (depending on the holding potential), and reversal potential as 0 mV. These parameters give good results for capacitance measurements in general. Check the "store" box for Cm, Gm, and Gs to save data.

17. The choice of recording time resolution depends on the purpose of the study. High time resolution is required to monitor fast surface area changes such as exocytosis, since it occurs in a subsecond time scale at nerve terminals. Although low time resolution is sufficient for clathrin-mediated endocytosis (which lasts tens of seconds or longer), high resolution recordings are often employed to capture those fast forms of endocytosis at better temporal resolution since nerve terminals display multiple forms of endocytosis (Fig. 2) [10, 39].

18. There are transient capacitance artefacts following depolarization (64 fF, τ = 227 ms for a 10 ms 0 mV pulse; 23 fF and τ = 96 ms for an action potential like pulse) [40, 41]. These capacitance jumps in both amplitude and kinetics remain intact after an effective block of exocytosis with botulinum toxin E, but it was sensitive to 0.25 mM Cd^{2+}. A SNARE-cleavage insensitive transient is also recorded in mossy fiber terminals [42]. The fast capacitance transient compromises the true capacitance changes from rapid endocytosis, particularly following a weak stimulation (e.g., a single action potential); it also complicate the interpretation of total exocytosis and the early phase of fast endocytosis component induced by a strong depolarization. To avoid this uncertainty, a short period of capacitance (200–500 ms) immediately following the depolarization is often simply omitted for analysis, depending on the stimulation and experimental purpose. Contamination may also arise from continuous exocytosis right after a strong stimulation since capacitance measure net surface area changes. The kinetics of fast endocytosis may be underestimated in this case.

19. Capacitance drift from unknown resources is often observed during capacitance recording from brain slices. Only Cm traces with no or minor drift are accepted for analysis. For endocytosis kinetics, the Cm traces with a minor, linear drift can be corrected by subtracting the baseline drift measured before stimulation. We used 5–10 s of baseline Cm to ensure a reliable baseline fitting [10]. Those capacitance traces with rapid, nonconstant drifts or large fluctuations are unreliable and rejected for analysis.

Acknowledgments

This work is partially supported by the National Institutes of Health (NIH) grants R01DK093953 (X.L.), 1R21NS101584-01 (X.L.), and the grant AAB1425-135-A5362 (X.L.). I thank Meyer Jackson for his comments on the manuscript.

References

1. Neher E, Marty A (1982) Discrete changes of cell membrane capacitance observed under conditions of enhanced secretion in bovine adrenal chromaffin cells. Proc Natl Acad Sci U S A 79(21):6712–6716

2. Gentet LJ, Stuart GJ, Clements JD (2000) Direct measurement of specific membrane capacitance in neurons. Biophys J 79(1):314–320

3. Gillis KD (2009) Techniques for membrane capacitance measurements. In: Bert Sakmann EN (ed) Single-Channel Reording, 2nd edn. Springer, pp 155–197

4. Lindau M, Fernandez JM (1986) IgE-mediated degranulation of mast cells does not require opening of ion channels. Nature 319(6049):150–153

5. von Gersdorff H, Matthews G (1994) Dynamics of synaptic vesicle fusion and membrane retrieval in synaptic terminals. Nature 367(6465):735–739

6. Heidelberger R, Heinemann C, Neher E, Matthews G (1994) Calcium dependence of the rate of exocytosis in a synaptic terminal. Nature 371(6497):513–515

7. Rieke F, Schwartz EA (1994) A cGMP-gated current can control exocytosis at cone synapses. Neuron 13(4):863–873

8. Beutner D, Voets T, Neher E, Moser T (2001) Calcium dependence of exocytosis and endocytosis at the cochlear inner hair cell afferent synapse. Neuron 29(3):681–690

9. Parsons TD, Lenzi D, Almers W, Roberts WM (1994) Calcium-triggered exocytosis and endocytosis in an isolated presynaptic cell: capacitance measurements in saccular hair cells. Neuron 13(4):875–883

10. Lou X, Paradise S, Ferguson SM, De Camilli P (2008) Selective saturation of slow endocytosis at a giant glutamatergic central synapse lacking dynamin 1. Proc Natl Acad Sci U S A

11. Schneggenburger R, Sakaba T, Neher E (2002) Vesicle pools and short-term synaptic depression: lessons from a large synapse. Trends Neurosci 25(4):206–212

12. Sun JY, Wu LG (2001) Fast kinetics of exocytosis revealed by simultaneous measurements of presynaptic capacitance and postsynaptic currents at a central synapse. Neuron 30(1):171–182

13. Lou X, Scheuss V, Schneggenburger R (2005) Allosteric modulation of the presynaptic Ca2+ sensor for vesicle fusion. Nature 435(7041):497–501

14. Hosoi N, Holt M, Sakaba T (2009) Calcium dependence of exo- and endocytotic coupling at a glutamatergic synapse. Neuron 63(2):216–229

15. Lin KH, Oleskevich S, Taschenberger H (2011) Presynaptic Ca2+ influx and vesicle exocytosis at the mouse endbulb of Held: a comparison of two auditory nerve terminals. J Physiol 589(Pt 17):4301–4320

16. Hsu SF, Jackson MB (1996) Rapid exocytosis and endocytosis in nerve terminals of the rat posterior pituitary. J Physiol 494(Pt 2):539–553

17. Hallermann S, Pawlu C, Jonas P, Heckmann MCP (2003) A large pool of releasable vesicles in a cortical glutamatergic synapse. Proc Natl Acad Sci U S A 100(15):8975–8980

18. Lindau M, Neher E (1988) Patch-clamp techniques for time-resolved capacitance measurements in single cells. Pflugers Arch 411(2):137–146

19. Sakmann B, Neher E (2009) Single-channel recording, 2nd edn. Springer, New York, NY, p xxii, 700

20. Gillis KD (2000) Admittance-based measurement of membrane capacitance using the EPC-9 patch-clamp amplifier. Pflugers Arch 439(5):655–664

21. Fan F et al (2015) Dynamin 2 regulates biphasic insulin secretion and plasma glucose homeostasis. J Clin Invest 125(11):4026–4041

22. Forsythe ID (1994) Direct patch recording from identified presynaptic terminals mediating glutamatergic EPSCs in the rat CNS, in vitro. J Physiol 479(Pt 3):381–387

23. Borst JG, Helmchen F, Sakmann B (1995) Pre- and postsynaptic whole-cell recordings in the medial nucleus of the trapezoid body of the rat. J Physiol 489(Pt 3):825–840

24. Kochubey O, Lou X, Schneggenburger R (2011) Regulation of transmitter release by Ca(2+) and synaptotagmin: insights from a large CNS synapse. Trends Neurosci 34(5):237–246

25. Borst JG, Soria van Hoeve J (2011) The calyx of Held synapse: from model synapse to auditory relay. Annu Rev Physiol

26. Korogod N, Lou X, Schneggenburger R (2007) Posttetanic potentiation critically depends on an enhanced Ca(2+) sensitivity of vesicle fusion mediated by presynaptic PKC. Proc Natl Acad Sci U S A 104(40):15923–15928

27. Wolfel M, Lou X, Schneggenburger R (2007) A mechanism intrinsic to the vesicle fusion machinery determines fast and slow transmitter release at a large CNS synapse. J Neurosci 27(12):3198–3210

28. Korogod N, Lou X, Schneggenburger R (2005) Presynaptic Ca2+ requirements and developmental regulation of posttetanic potentiation at the calyx of Held. J Neurosci 25(21):5127–5137

29. Lou X, Korogod N, Brose N, Schneggenburger R (2008) Phorbol esters modulate spontaneous and Ca2+-evoked transmitter release via acting on both Munc13 and protein kinase C. J Neurosci 28(33):8257–8267

30. Dulubova I et al (2005) A Munc13/RIM/Rab3 tripartite complex: from priming to plasticity? EMBO J 24(16):2839–2850

31. Mahapatra S, Fan F, Lou X (2016) Tissue-specific dynamin-1 deletion at the calyx of Held decreases short-term depression through a mechanism distinct from vesicle resupply. Proc Natl Acad Sci U S A 113(22):E3150–E3158

32. Mahapatra S, Lou X (2016) Dynamin-1 deletion enhances post-tetanic potentiation and quantal size after tetanic stimulation at the calyx of Held. J Physiol

33. Fan F, Funk L, Lou X (2016) Dynamin 1- and 3-mediated endocytosis is essential for the development of a large central synapse in vivo. J Neurosci 36(22):6097–6115

34. Debus K, Hartmann J, Kilic G, Lindau M (1995) Influence of conductance changes on patch clamp capacitance measurements using a lock-in amplifier and limitations of the phase tracking technique. Biophys J 69(6):2808–2822

35. Horrigan FT, Bookman RJ (1994) Releasable pools and the kinetics of exocytosis in adrenal chromaffin cells. Neuron 13(5):1119–1129

36. Kilic G, Lindau M (2001) Voltage-dependent membrane capacitance in rat pituitary nerve terminals due to gating currents. Biophys J 80(3):1220–1229

37. Wolfel M, Schneggenburger R (2003) Presynaptic capacitance measurements and Ca2+ uncaging reveal submillisecond exocytosis kinetics and characterize the Ca2+ sensitivity of vesicle pool depletion at a fast CNS synapse. J Neurosci 23(18):7059–7068

38. Sun JY et al (2004) Capacitance measurements at the calyx of Held in the medial nucleus of the trapezoid body. J Neurosci Methods 134(2):121–131

39. Saheki Y, De Camilli P (2012) Synaptic vesicle endocytosis. Cold Spring Harb Perspect Biol 4(9):a005645

40. Yamashita T, Hige T, Takahashi T (2005) Vesicle endocytosis requires dynamin-dependent GTP hydrolysis at a fast CNS synapse. Science 307(5706):124–127

41. Wu W, Xu J, Wu XS, Wu LG (2005) Activity-dependent acceleration of endocytosis at a central synapse. J Neurosci 25(50):11676–11683

42. Delvendahl I, Vyleta NP, von Gersdorff H, Hallermann S (2016) Fast, temperature-sensitive and clathrin-independent endocytosis at central synapses. Neuron 90(3):492–498

Chapter 9

Assaying Mutants of Clathrin-Mediated Endocytosis in the Fly Eye

Elsa Lauwers and Patrik Verstreken

Abstract

Clathrin-mediated endocytosis plays essential roles both during and after development, and loss-of-function mutants affected in this process are mostly not viable. Different approaches have been developed to circumvent this limitation, including resorting to mosaic model organisms. We here describe the use of FLP/FRT-mediated mitotic recombination to generate *Drosophila melanogaster* having homozygous mutant eyes while the rest of their body is heterozygous. We then present a detailed protocol for assessing the consequences of these loss-of-function mutations on endocytosis in the photoreceptors of living fruit flies by recording electroretinograms.

Key words Drosophila, Endocytosis, Mitotic recombination, Genetic mosaic, FLP/FRT system, Electroretinogram

1 Introduction

Loss-of-function mutations of genes encoding endocytic factors often cause early lethality [1] or complex phenotypes resulting from defects accumulation during development [2]. Several alternatives have thus been developed, including pharmacological approaches [3] or acute protein inactivation methods [4–6]. In Drosophila, another popular tool is the generation of genetic mosaics by means of FLP/FRT mitotic recombination [7]. Since fruit flies possess well-developed compound eyes [8] and can live and mate without vision, this site-specific recombination system has notably been adapted to produce homozygous mutant eye tissue in an otherwise wild-type genetic background with high efficiency (Fig. 1) [9, 10]. In this system flippase (FLP)-mediated recombination takes place between two flippase recognition target (FRT) sites positioned at identical loci and in the same orientation

---

The original version of this chapter was revised. A correction to this chapter can be found at https://doi.org/10.1007/978-1-4939-8719-1_19

Laura E. Swan (ed.), *Clathrin-Mediated Endocytosis: Methods and Protocols*, Methods in Molecular Biology, vol. 1847, https://doi.org/10.1007/978-1-4939-8719-1_9, © Springer Science+Business Media, LLC, part of Springer Nature 2018

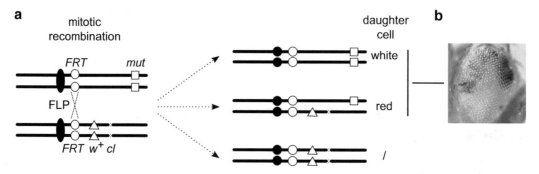

Fig. 1 Generating homozygous mutant fly eyes using the FLP/FRT recombination system. (*a*) Schematic representation of FLP-mediated recombination between two chromosome arms carrying *FRT* sites using the *eyFLP* system used here. (*b*) Typical aspect of mosaic eyes obtained with the *eyFLP* system shown in (*a*)

on homologous chromosomes. When it occurs after DNA replication, recombination makes the region that lies distal to the FRT site homozygous in theoretically 50% of the daughter cells (Fig. 1). By placing the *FLP* cDNA under transcriptional control of an eye-specific enhancer from the *eyeless* (*ey*) gene, mitotic recombination is restricted to the developing visual system. Note that a lower level of *FLP* expression is also detected in neurons of the optic lobes of the brain [9] and that more eye specific promoters have been introduced [11]. Recombination events in the eye can be traced in a *white* mutant (*w⁻*) background by means of a *white⁺* (*w⁺*) marker located distal to the FRT site (Fig. 1a). To further increase the size of the *w⁻* clone (i.e., have more than 50% of mutant eye cells), a recessive cell lethal (*cl*) mutation has been placed onto the marked *FRT w⁺* chromosome arm so that twin-spot wild-type clone cells carrying two *w⁺* alleles are eliminated (Fig. 1) [9].

By itself the *eyFLP/FRT* system does not cause defects in mosaic eye size, morphology or physiology [9, 10], allowing to study the effect of a mutation of interest in the fly eye and to perform forward genetic screens based on an eye phenotype. The most commonly used readout for eye function is the electroretinogram (ERG). While ERGs have historically been applied to study phototransduction [12, 13], they also constitute a robust assay to assess neuronal communication between photoreceptors and second-order brain neurons; this communication heavily relies on endocytosis [14, 15]. In this protocol we describe how to record and interpret ERGs from mosaic Drosophila eyes carrying mutations in components of the clathrin-mediated endocytosis (CME) machinery. We illustrate our point with a severe loss-of-function allele of the *synaptojanin* gene on the 2R chromosome arm [14]. It should be noted that numerous other fly lines carrying *eyFLP* and *FRT* transgenes at different insertion sites, thus allowing the study of mutations on different chromosome arms, are also available (http://flystocks.bio.indiana.edu/Browse/misc-browse/flp.php).

2   Materials

All solutions are prepared using ultrapure water (18 MΩ cm at 25 °C). All reagents and solutions are stored at room temperature.

2.1 Fly Genetics

1. Laboratory equipment required for Drosophila husbandry, including incubators, microscopes, standard cornmeal, molasses, and yeast fly food (http://flystocks.bio.indiana.edu/Fly_Work/media-recipes/molassesfood.htm), and experience with simple genetic crosses.

2. Flies with *eyFLP* on the X chromosome, FRT, w⁺ and a recessive cell lethal mutation on the 2R chromosome arm (y^{d2} w^{1118} $P\{ry^{+t7.2}$ = ey-$FLP.N\}2$ $P\{GMR$-$lacZ.C(38.1)\}TPN1;$ $P\{ry^{+t7.2}$ = $neoFRT\}42D$ $P\{w^{+t*}$ ry^{+t*} = $white$-$un1\}47A$ $l(2)$ cl-$R11^1/CyO$, y^+), available from the Bloomington Stock Center (stock number: 5617) [9]. From here on we will refer to these flies as *yw eyFLP; FRT42D w + cl/CyO*.

3. Flies with *eyFLP* on the X chromosome and the mutation of interest on the *FRT42D* chromosome arm (y^{d2} w^{1118} $P\{ry^{+t7.2}$ = ey-$FLP.N\}2$ $P\{GMR$-$lacZ.C(38.1)\}TPN1;$ $P\{ry^{+t7.2}$ = $neoFRT\}42D$ $synj^1/CyO$-$KrGFP$). From here on we will refer to these flies as *yw eyFLP; FRT42D synj¹/CyO*. These flies were originally obtained by EMS mutagenesis of isogenized *y w/Y*; $P\{ry^{+t7.2}$ = $neoFRT\}42D$ males [14]. The procedure to recombine a given mutation with the appropriate *FRT* insertion is described on the Bloomington Stock Center website (http://flystocks.bio.indiana.edu/Browse/misc-browse/FRT_protocol_Carthew.html).

2.2 Electro-retinogram Setup

1. Benchtop Faraday cage (750 mm wide and 1200 mm high), with a breadboard with ferromagnetic stainless steel top skin, a stereomicroscope and a white LED lamp as a source for the light flashes.

2. Cold light source to visualize the fly while adjusting the electrodes.

3. Amplifier with a headstage, stimulator, and digitizer.

4. Hook-up wire (1.65 mm diameter), three BNC cables, a BNC connector, a BNC to dual banana plug adapter, three cables with banana plugs, mini-USB cable, cutting pliers, and cable stripper.

5. 0.25 mm silver wire, petri dish, 3 M KCl solution, laboratory power supply, and lead cables ending with small alligator clips to coat the wire with chloride.

6. Soldering station and solder (0.71 mm wire).

7. Electrode holders for the reference electrode and for the measuring electrode (extracellular electrode holder with silver wire). Ensure that the electrode holder size matches the outside diameter of the capillaries used to make the glass electrodes (see below).

8. Two micromanipulators and two magnetic stands.

9. Personal computer (PC) with the Axoscope 10.5 software (Molecular Devices) as an interface.

10. Optional: a light-blocking curtain entirely covering the Faraday cage is necessary if the setup is installed in a room where the lights cannot be switched off. This can be custom-made from light-blocking fabric and if desired mounted on a curtain rail for easier access. All supplies can be found in any home decor store.

2.3 Electrodes Preparation

1. Borosilicate glass capillaries (1.5 mm with filament, outer diameter: 1.5 mm thickness: 0.315 mm) and laser-based micropipette puller to produce glass electrodes. These can be stored in a petri dish on two bars of modeling clay.

2. 3 M NaCl solution, 1 ml syringe, and 28 gauge needle to backfill the glass electrodes.

2.4 ERG Recording

1. Microscope slides and glue (glue must be a transparent liquid such as Pritt universal glue, Henkel, or similar) to immobilize flies.

2. CO_2-equipped fly station consisting of a stereomicroscope with a cold light source and a fly pad with frame, connected via tubing and a foot valve to a source of CO_2.

3. Fine forceps to manipulate the flies.

2.5 ERG Profile Quantification

1. Personal Computer (PC) equipped with Clampfit 10.5 software (Molecular Devices) to analyze ERG traces and a spreadsheet software to collect and organize the data.

2. Optional: software (e.g., GraphPad Prism 6 and Inkscape 0.91) to generate graphs and figures.

3   Methods

All procedures are carried out at room temperature, unless indicated otherwise. Note that the steps of Subheadings 3.2 and 3.4 only need to be performed once, when the ERG setup is installed (*see* **Note 1**).

3.1 Fly Genetics

1. Make sure to amplify the two parent fly lines at least 3 weeks before the experiments are scheduled (*see* **Note 2**).

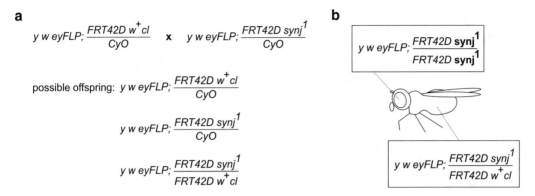

Fig. 2 Crossing scheme. (*a*) A cross between the two parent fly lines described in Subheading 2.1 gives rise to offspring of three possible genotypes. (*b*) Because of FLP-mediated recombination, flies that carry the two 2R chromosome arms with an ***FRT42D*** site have a different genotype in the eye than in the rest of their body

2. About 12 days before experiments start, collect virgin female flies from one parent line and cross them to males from the other line (Fig. 2a). Incubate the flies at 20 °C.

3. When adult flies emerge, collect those that do not carry any CyO balancer chromosome—i.e., those that have straight, noncurled wings [16] (Fig. 2a). The genotype of these flies is *yw eyFLP; FRT42D cl w⁺/FRT42D synj¹*, except for their eyes where most cells are homozygous for *synj¹* (Fig. 2b). The eyes thus appear mostly unpigmented (w^-) with few red cells (w^+) that are heterozygous for the 2R chromosome arm. Avoid recording ERGs from flies that are just eclosed and keep those at 20 °C for at least a day before using them for experiments (*see* **Note 3**).

3.2 Installing the ERG Setup

1. A schematic representation of the ERG setup is depicted in Fig. 3a. Place the Faraday cage on a table and install the breadboard, the stereomicroscope and the LED lamp in the cage. We typically tape the LED lamp to an empty small fly food vial and place it on the stereomicroscope at the appropriate height to directly illuminate the fly. Place the PC, amplifier, stimulator, digitizer, and LCD lamp outside the cage on the same table. Connect the digitizer to the output port of the amplifier (channel 0) and to the auxiliary output port of the stimulator (channel 1) using BNC cables, and to the PC via a mini-USB cable. Connect the LED lamp to the main output port of the stimulator via two cables with banana plugs, each soldered to one of the lamp pins, a BNC with dual banana plug adapter, a BNC connector, and a BNC cable. Plug a banana cable to the ground (chassis) port of the amplifier and wrap the bare cable end to a metallic element of the stereomicroscope (e.g., a screw).

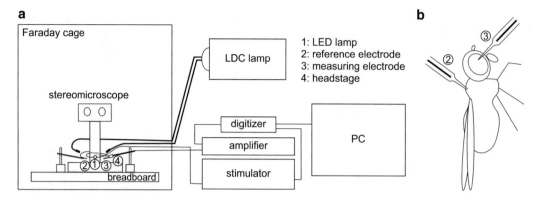

Fig. 3 Electroretinogram setup. (**a**) Schematic representation of our ERG setup. (**b**) Diagram showing an immobilized fly and the position of the reference and measuring electrodes

2. Coat an 8 cm piece of sliver wire with chloride by electrolysis: Connect the wire to the power supply by clamping the extremities of the wire in the jaws of the alligator clips ending the lead cables. Dip the rest of the wires into a petri dish filled with 3 M KCl, making sure that the clips and the lead cables do not come in contact with the solution. Apply a 5 mV current for 5 min. The solution near the wire will bubble and the wire will progressively turn darker or brownish. After 5 min remove the wire from the clips and rinse it with deionized water. These chlorated silver electrodes need to be replaced periodically, when the silver wire is again exposed and the quality of the ERG recordings is thus not satisfying anymore (*see* **Note 4**).

3. Build the reference electrode by soldering one piece of chlorated silver wire to a 30 cm piece of hook-up wire, and solder the other extremity of this wire to the minus input pin of the amplifier headstage. Tape the silver wire to the reference electrode holder, leaving about 2 cm of free wire at the extremity. Place the reference electrode holder in a micromanipulator, itself mounted on a magnetic stand, at the left-hand side of the stereomicroscope.

4. Connect the extracellular electrode holder to the positive input pin of the amplifier headstage. Place the headstage in a micromanipulator, itself mounted on a magnetic stand, at the right-hand side of the stereomicroscope. Connect the headstage to the input port of the amplifier.

3.3 Preparing Glass Electrodes

1. Adjust the settings of the puller in order to produce two sharp, tapered electrodes from a glass capillary; these settings depend on the puller and the capillaries used. Electrodes can be stored in a petri dish on a bar of modeling clay.

2. Use a 1 ml syringe with a MicroFil needle to backfill the electrodes with 3 M NaCl. Make sure there is no air bubble in the

glass electrodes to ensure proper signal transmission and accurate measurements. Place the filled glass electrode in the reference and measuring electrode holders, inserting the silver wire in the glass capillary, and secure them by tightening the metal or Plexiglas screw, respectively.

3.4 Configuring the Amplifier and Stimulator

1. Set the amplifier in the AC mode by selecting a cut off frequency of 0.1 Hz using the high pass filter control. Set the low pass filter control to 300 Hz and the gain to 100.

2. Set the stimulator settings as follows. Period: 1 s, pulse width: 500 ms, pulse delay: 500 ms, mode: run, amplitude: 7. This amplitude is generally suitable to obtain a clear ERG response from flies with nonpigmented eyes. If the amplitude of photoreceptor depolarization (*see* Subheading 3.6) appears very small, typically under 5 mV, the stimulation amplitude may need to be increased further.

3. Use the Axoscope 10.5 software to create a stimulation protocol (Use the "Acquire" menu, then "New Protocol"). Select the "fixed-length events" acquisition mode, and a duration of 20 s. Record 2000 samples per s (sampling rate per signal: 1000 Hz, sweep length per event per signal: 2). Use the IN1 port as trigger source and define a pretrigger length of 500 ms and a trigger threshold of 4A. Set the polarity as "rising." Save the protocol and name it for example "ERG test."

3.5 Recording ERGs

1. Place the flies selected in Subheading 3.1 in the dark about 10 min before starting the experiment. Dark adaptation increases the amplitude of photoreceptor depolarization during light pulses (*see* Subheading 3.6).

2. Turn the PC, the stimulator, the amplifier and the LCD lamp on. Open the Axoscope 10.5 software and use the "Acquire" menu and then "Open Protocol" to open the "ERG test" protocol created in Subheading 3.4. Use the "Configure" menu then "Digitizer" to scan until the software finds the mini Digi 1A digitizer, and select this digitizer for further use. Finally, go the "File" menu and then "Set Data File Names" to select the name that will be given to the files and the location where they should be saved. Note that by default the software will generate a new file with an extension starting at 000 and incremented by 1 for every ERG recording.

3. Prepare a microscope slide with immobilized flies to record ERGs from. Working at the fly station, use a regular pipette tip to dispense small drops of liquid glue on a slide, leaving about 5 mm between the drops. Anaesthetize the flies with CO_2 and carefully pick them up one at a time by their wings with the forceps. Place each fly on its side on a drop of glue so that it lies

with its back oriented toward the left (Fig. 3b). Make sure to place the wings and the legs in the glue and to glue the head tightly while leaving the proboscis free (*see* **Note 5**).

4. Place the slide with the glued flies under the stereomicroscope in the Faraday cage. Turn on the LCD lamp and use the micromanipulators to bring the electrodes to a position close to the first immobilized fly to record ERGs from. Focusing on the eye, make sure that the tip of both electrodes is clearly visible. Carefully insert the tip of the reference electrode into the thorax of the fly (Fig. 3b), going through the cuticle without pushing the electrode too deep. Inserting the electrode too deep in the thorax typically adds noise to the ERG signal. When detecting noise during recordings, it is often useful to pull back the reference electrode a little bit. When the reference electrode is in place, adjust the measuring electrode on the surface of the eye (Fig. 3b), with just enough pressure to cause the eye to form a small dimple without penetrating the cornea. Close the shutter of the LCD lamp and, if needed, close the curtain to make sure that the LED lamp is the only source of light.
5. Using the Axoscope 10.5 software, first press the "play button" to make sure that a proper, noise-free ERG signal is generated. Then press the "record button" (or go to the "Acquire" menu and then "Record"). The "ERG test" protocol will generate 5 pulses of light before stopping. The resulting ERG traces that appear on the screen will automatically be saved in the folder and with the files names that were defined as described above. Make sure to keep a record, e.g., in an Excel file, of the genotype of each fly and the corresponding ERG data file.

3.6 Interpreting and Quantifying ERG Profiles

1. Open the ERG data files with the Clampfit 10.5 software. The ERG traces that were recorded will be displayed on top of each other. These traces show three different components, namely photoreceptors depolarization and ON- and OFF-transients at the onset and offset of the light pulse, respectively (Fig. 4a). Since ERGs are extracellular recordings, depolarization of the photoreceptors manifests as a more negative potential that progressively returns to baseline after the light pulse. This depolarization results from a complex phototransduction cascade [8] and indicates that the light has been perceived properly. In contrast, ON- and OFF-transients arise from brain neurons in the lamina that receive synaptic input from the photoreceptors [13]. Any mutation that disturbs the synchronicity of neuronal communication between photoreceptors and lamina neurons will thus cause a reduction in the amplitude of the ON and OFF peaks. These results must be interpreted with

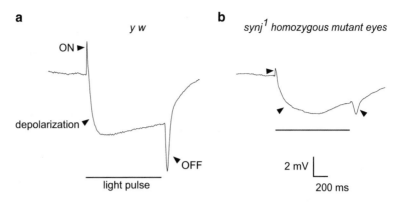

Fig. 4 A ***synj^1*** homozygous mutation severely affects neurotransmission. (***a***) Different components of a typical ERG trace recorded from a control animal (in this case a ***yw*** fly). The ON, OFF, and depolarization are indicated on the graph, ***see*** the Subheading 3.6 for details. (***b***) Representative ERG trace from a fly having homozygous ***synj^1*** eyes and a heterozygous ***synj^1/+*** body. Note the reduced amplitude of ON and OFF transients

caution as there can be multiple causes for reduced ON and OFF amplitude. One of these causes can be defective synaptic vesicle recycling caused by a defect in CME, as observed for example in flies with *synaptojanin* mutant eyes [14] (Fig. 4b).

2. The amplitude of these different ERG components can easily be quantified. If needed, it is possible to first manually select the traces that would display too much noise and delete them using the "Delete" button on the keyboard. Go to the "Analyze" menu and click "average traces" to obtain a single average ERG trace from the 5 traces that were acquired during one recording. Before measuring the different amplitudes the baseline of the trace must be adjusted to be 0 mV. To do so place the cursor 1 at position 0 along the x (time) axis, and the cursor 2 at position 499 ms, just before the start of the light pulse. This can be done by dragging the cursors with the mouse pointer, or by double-clicking on each cursor to adjust its properties. Then go to the "Analyze" menu again and click "Adjust." Select "baseline" from the drop-down menu and use the dialog box to subtract from each point of the trace the average potential value from all the points between cursors 1 and 2. Make sure that the following properties are selected in the dialog box that opens when double-clicking on cursor: visible, text box with acquisition time, data value and unit. Move cursor 1 along the trace and read the potential value at the maximum of the ON peak (ON amplitude), at time 990 ms—just before the end of the light pulse, when photoreceptor depolarization has stabilized—(depolarization amplitude) and at the minimum of the OFF peak (OFF amplitude). Compile these data in a spreadsheet software such as Excel.

3. Use Excel or GraphPad Prism 6 software to create graphs with the collected data. It may also be desirable to generate a figure showing a representative ERG trace. This can for example be done by exporting all the data points (after averaging different traces or not) to Excel and then generating a line graph. To do this, go to the "Edit" menu in Clampfit and click on "transfer traces." In the dialog box that opens select "result window" as destination window, and "full trace" as region to transfer, and tick the "start time base at zero" box. For trace selection, click on select and choose "specified signal of IN0 (mV)" and "all visible traces," then click on the OK button twice. Finally copy the data that appears in the result window and then paste it in an Excel file (*see* **Note 6**).

4   Notes

1. While the FLP/FRT system is the most widely used site-specific recombination system in *Drosophila* several other options exist, e.g., the Cre/loxP system. Available fly lines are listed on the Bloomington Stock Center webpage (http://flystocks.bio.indiana.edu/Browse/misc-browse/recomb_alt.php).

2. It is important to make sure that the food contains enough carotenoid. Carotenoid serves as a precursor for 3-hydroxy retinal, i.e., the chromophore for rhodopsin [17]. The standard fly food described in the above article contains enough carotenoid from the zeaxanthin contained in the corn molasses; if necessary beta-carotene can be added at a concentration of 125 mg/ml to maximize visual pigment and visual sensitivity [18].

3. The methods we describe here can be used to perform forward genetic screens [14]. For this purpose, male flies carrying an *FRT* insertion on the chromosome arm of choice should first be mutagenized. These males can then be crossed with virgin female flies carrying both an *eyFLP* transgene as well as an *FRT* site inserted at the same position and flanked by a w^+ marker and a *cl* mutation. The resulting flies will have mosaic eyes, where the majority of photoreceptors are homozygous for the induced mutations.

4. As an alternative to the procedure described in Subheading 3.2, it is also possible to chlorate silver wire by simple immersion in common household bleach overnight. After this treatment, the wires should be thoroughly rinsed with deionized water.

5. It is advisable to immobilize a limited number of flies, and to record ERGs within 10 min after the flies have been glued.

Note that ERGs should be recorded from a few wild-type flies at the start of every experiment, as well as every time the glass electrodes and/or the silver wire are replaced, to make sure the setup is working properly.

6. A figure showing an ERG trace can also be generated by copying the graphical data from Clampfit as an Axon metafile and pasting it in a graphical software, e.g., Inkscape 0.91. To do this, click on the graph and press "Ctrl + C". Choose the "Copy Analysis Window to clipboard as a metafile" and simply paste this data in a new Inkscape document. Note that it is essential to make sure that the different traces are displayed at the same scale before deleting undesired component of the image, e.g., the different axis and gridlines.

References

1. Bazinet C, Katzen AL, Morgan M et al (1993) The drosophila clathrin heavy chain gene: clathrin function is essential in a multicellular organism. Genetics 134(4):1119–1134

2. Overstreet E, Fitch E, Fischer JA (2004) Fat facets and liquid facets promote delta endocytosis and delta signaling in the signaling cells. Development 131(21):5355–5366

3. von Kleist L, Haucke V (2011) At the crossroads of chemistry and cell biology: inhibiting membrane traffic by small molecules. Traffic 13(4):1–10

4. Kasprowicz J, Kuenen S, Miskiewicz K et al (2008) Inactivation of clathrin heavy chain inhibits synaptic recycling but allows bulk membrane uptake. J Cell Biol 182(5):1007–1016

5. Marek KW, Davis GW, Francisco S (2002) Transgenically encoded protein photoinactivation (FlasH-FALI): acute inactivation of Synaptotagmin I. Neuron 36:805–813

6. Kasprowicz J, Kuenen S, Swerts J et al (2014) Dynamin photoinactivation blocks Clathrin and α-adaptin recruitment and induces bulk membrane retrieval. J Cell Biol 204(7):1141–1156

7. Xu T, Rubin GM (1993) Analysis of genetic mosaics in developing and adult drosophila tissues. Development 1237:1223–1237

8. Montell C (1999) Visual transduction in drosophila. Annu Rev Cell Dev Biol 15:231–268

9. Newsome TP, Asling B, Dickson BJ (2000) Analysis of drosophila photoreceptor axon guidance in eye-specific mosaics. Development 127(4):851–860

10. Stowers RS, Schwarz TL (1999) A genetic method for generating drosophila eyes composed exclusively of mitotic clones of a single genotype. Genetics 152:1631–1639

11. Bazigou E, Apitz H, Johansson J et al (2007) Anterograde jelly belly and Alk receptor tyrosine kinase signaling mediates retinal axon targeting in *Drosophila*. Cell 128(5):961–975

12. Pak WL (1995) Drosophila in vision research. The Friedenwald lecture. Invest Ophthalmol Vis Sci 36(12):2340–2357

13. Heisenberg M (1971) Separation of receptor and lamina potentials in the electroretinogram of normal and mutant drosophila. J Exp Biol 55(1):85–100

14. Verstreken P, Koh T-W, Schulze KL et al (2003) Synaptojanin is recruited by endophilin to promote synaptic vesicle uncoating. Neuron 40(4):733–748

15. Fabian-Fine R, Verstreken P, Hiesinger PR et al (2003) Endophilin promotes a late step in endocytosis at glial invaginations in drosophila photoreceptor terminals. J Neurosci 23(33):10732–10744

16. Chyb S, Gompel N (eds) (2013) Atlas of drosophila morphology. Academic Press, London

17. Gu G, Yang J, Mitchell KA, O'Tousa JE (2004) Drosophila ninaB and ninaD act outside of retina to produce rhodopsin chromophore. J Biol Chem 279(18):18608–18613

18. Picking WL, Chen DM, Lee RD et al (1996) Control of drosophila opsin gene expression by carotenoids and retinoic acid: northern and western analyses. Exp Eye Res 63(5):493–500

Chapter 10

Reconstitution of Clathrin Coat Disassembly for Fluorescence Microscopy and Single-Molecule Analysis

Till Böcking, Srigokul Upadhyayula, Iris Rapoport, Benjamin R. Capraro, and Tom Kirchhausen

Abstract

The disassembly of the clathrin lattice surrounding coated vesicles is the obligatory last step in their life cycle. It is mediated by the coordinated recruitment of auxilin and Hsc70, an ATP-driven molecular clamp. Here, we describe the preparation of reagents and the single-particle fluorescence microscopy imaging assay in which we visualize directly the Hsc70-driven uncoating of synthetic clathrin coats or clathrin-coated vesicles.

Key words Clathrin-coated vesicle, Auxilin, Microfluidics, TIRF microscopy, Clathrin uncoating

1 Introduction

The disassembly of the clathrin coat occurs shortly after the scission of the coated vesicle from the donor membrane and represents a key step in the life cycle of a clathrin-coated vesicle (CCV) [1–4]. The uncoating reaction depends on ATP hydrolysis [5–7], and is jointly mediated by interactions by the molecular chaperone heat shock cognate protein 70 (Hsc70) that binds to the unstructured C-terminal tail of clathrin heavy chain with its cochaperone auxilin [8] also recruited to the coated vesicle [3, 5, 9]. Ensemble in vitro measurements of the uncoating reaction of natural coated vesicles or clathrin coats (CCs) assembled from purified clathrin and adaptor proteins (AP-2) have resolved key molecular steps of the process [5, 7]. We recently developed a fluorescence microscopy method to observe the uncoating reaction at the level of individual synthetic clathrin coats reconstituted with the AP-2 adaptor complex [8]. Here, we describe in detail use of this method to

The original version of this chapter was revised. A correction to this chapter can be found at https://doi.org/10.1007/978-1-4939-8719-1_19

Till Böcking and Srigokul Upadhyayula contributed equally to the chapter.

Laura E. Swan (ed.), *Clathrin-Mediated Endocytosis: Methods and Protocols*, Methods in Molecular Biology, vol. 1847, https://doi.org/10.1007/978-1-4939-8719-1_10, © Springer Science+Business Media, LLC, part of Springer Nature 2018

follow the uncoating mediated by recombinant Hsc70 and auxilin of fluorescently tagged CCs or of in vitro reconstituted CCVs. Fluorescent clathrin trimers reconstituted from recombinant heavy chain and light chain labeled with a fluorophore were either mixed with the adaptor protein complex AP-2 for assembly into CCs (Fig. 1a) or mixed with AP-2 and extruded liposomes with tyrosine-motif (YQRL) containing peptidolipid for assembly into CCVs (Fig. 1b). The clathrin-coated structures were immobilized at the bottom of a microfluidic channel device on a passivated glass coverslip coated with a monoclonal antibody specific for clathrin light chain A (Fig. 2) and appeared as bright diffraction-limited spots in the clathrin channel. In the case of reconstituted CCVs, the signal from the lipid vesicle stained with membrane-binding dyes was detected at the locations corresponding to immobilized CCVs, whereby different color dyes were used to distinguish vesicles made with different lipid compositions (Fig. 1c). ATP-loaded Hsc70 labeled with a different fluorophore together with auxilin were then injected into the microfluidic channel to initiate the uncoating reaction (Fig. 2); the time course of the fluorescence signals associated with single clathrin coats and Hsc70 are then recorded using multiwavelength time-lapse total internal reflection fluorescence (TIRF) microscopy (Fig. 2). Traces of clathrin intensity were extracted from the fluorescence movies at each location corresponding to a single coat, whereby the fluorescence intensity was converted into number of molecules using a calibration value determined by single-molecule photobleaching. The resulting clathrin traces showed the process of clathrin uncoating for individual clathrin structures (Fig. 3). Likewise, traces of Hsc70

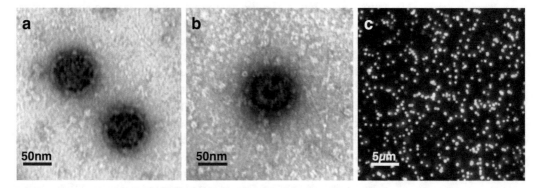

Fig. 1 Transmission electron microscopy (TEM) and total internal reflection (TIRF) images of in vitro assembled clathrin coats and clathrin-coated vesicles. (**a**) Negative stain TEM images of clathrin coats (CCs) and (**b**) a clathrin coated vesicle (CCV). (**c**) Multiplexed visualization of CCV components using TIRF microscopy. Empty clathrin coats are shown in green, clathrin coats surrounding liposomes containing PI(3)P, PI(4,5)P$_2$, DOPC, and YQRL containing peptidolipid are shown in orange, and clathrin coats surrounding liposomes containing PI(5) P, PI(4,5)P$_2$, DOPC, and YQRL containing peptidolipid are shown in cyan. The clathrin light chain A is labeled with Alexa Fluor 488 (visualized in green color), PI(3)P and PI(5)P containing liposomes are labeled with DiI and DiD lipid dyes (visualized in red and blue colors), respectively

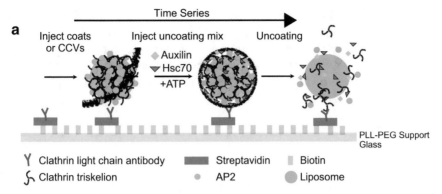

Fig. 2 Schematic of surface chemistry and uncoating assay. Glass coverslips are coated with a layer of a copolymer composed of poly-L-lysine and biotinylated poly(ethylene glycol) (PLL-PEG) to minimize nonspecific adsorption of proteins to the surface. The surface is then modified with streptavidin followed by the biotinylated monoclonal antibody CVC.6 directed against mutated rat clathrin light chain A. In vitro assembled CCs clathrin coats or CCVs are then captured onto the surface via the antibody (shown here for CCV). Uncoating reagents (Hsc70, auxilin, ATP) are then added to the immobilized clathrin coats or CCV to initiate the uncoating reaction. Different components of the reaction can be labeled with spectrally distinct fluorophores to allow observation of their arrival and/or departure at the locations of individual coats/CCVs by multiwavelength TIRF microscopy

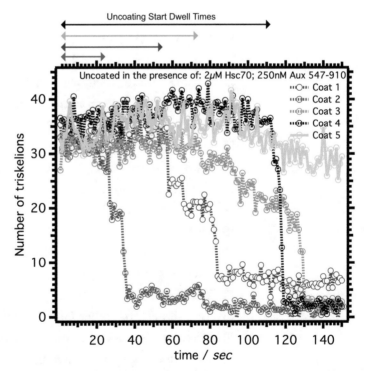

Fig. 3 Kinetics of clathrin-coat disassembly. Single clathrin-coat disassembly tracings after injecting ΔPTEN-Auxilin, Hsc70 with ATP. The stochastic behavior illustrated by differences in the dwell time to start the clathrin lattice disassembly (monitored via labeled clathrin light chain A signal) is shown for five different traces

association with the coats were determined from the colocalized signals of Hsc70 at the location of each clathrin coat. The loss of clathrin fluorescence and amount of recruited Hsc70 were determined from several hundred coats imaged in the field of view allowing us to extract kinetic parameters of Hsc70 binding and clathrin uncoating [8].

2 Materials

2.1 Equipment

1. Shaker for baculovirus cultures (28C, dark).
2. Sonicator.
3. Metal beaker.
4. Bacterial shaker.
5. FPLC.
6. Ultracentrifuge Sorvall or equivalent.
7. Benchtop centrifuge for pelleting bacterial/baculovirus cultures.
8. Spectrophotometer.
9. Tissue grinder (e.g., Potter Homogenizer or similar).
10. 0.2 µm syringe filters.
11. Baffled flask for bacterial culture.
12. Spinner flask for baculovirus culture.
13. SDS-PAGE gel electrophoresis setup.
14. Kitchen blender.
15. Argon-flow sample dryer.
16. Electron microscope.
17. Vacuum desiccator.
18. Laboratory oven (70 °C).
19. Hot plate.
20. Headway spin coater (or similar).
21. Laurell Spin Coater (or similar).
22. THINKY ARE-250 Mixer.
23. Technics Plasma etcher 500-II (or similar).
24. Micropuncher to make holes in PDMS.
25. Glow discharge plasma chamber.
26. Access to clean-room facility.

2.2 Reagents for Expression and Purification of Clathrin Heavy Chain

1. pFastBac-based plasmid encoding the heavy chain [10].

2. DH10Bac bacterial cells, and Sf9 and High Five insect cells (ThermoFisher) (Ref. 10).

3. LB agar containing 7 μg/mL gentamycin, 10 μg/mL tetracycline, 50 μg/mL kanamycin, 100 μg/mL Bluo-Gal, and 40 μg/mL isopropyl β-D-1-thiogalactopyranoside (IPTG).

4. Sf900-II (Gibco/ThermoFisher), and serum-free medium for insect cell culture (either Excell 420 (Sigma-Aldrich) [10] or alternatively Express Five (Gibco/ThermoFisher).

5. Lysis Buffer 1: 50 mM Tris–HCl pH 8.0, 300 mM NaCl, 0.5 mM dithiothreitol (DTT), 1 mM ethylenediaminetetraacetic acid (EDTA), 0.05 mg/mL RNase A.

6. Cage Formation Buffer: 20 mM Mes pH 6.2, 2 mM CaCl$_2$, 0.02% NaN$_3$, 0.5 mM DTT.

7. Buffer A: 50 mM Mes pH 6.5, 100 mM NaCl, 1 mM EGTA, 0.5 mM MgCl$_2$, 0.02% NaN$_3$, 0.5 mM DTT.

8. Buffer B: 2.4 M Tris–HCl pH 7.4, 0.04% NaN$_3$, 1 mM DTT.

9. Column Buffer 1(filter before use): 0.5 M Tris–HCl pH 7.4, 0.02% NaN$_3$, 0.5 mM DTT.

10. Low phosphate buffer (filter before use): 10 mM NaH$_2$PO$_4$ pH 7.1, 100 mM NaCl, 0.02 NaN$_3$, 0.5 mM DTT.

11. High phosphate buffer (filter before use): 500 mM NaH$_2$PO$_4$ pH 7.1, 100 mM NaCl, 0.02 NaN$_3$, 0.5 mM DTT.

12. Alkaline Buffer: 200 mM NaOH with 1% SDS.

13. Isopropanol.

14. Tris–EDTA buffer: 10 mM Tris–HCl, pH 7.5, 1 mM EDTA.

15. PMSF (stock concentration) or PEFA (stock concentration).

16. 0.5 M EDTA stock solution, pH 8.0.

17. HiLoad 26/600 Superdex 200 gel filtration column (GE Life Sciences).

18. Econo-Pac CHT-II Ceramic Hydroxyapatite column, 5 mL (Bio-Rad).

2.3 Reagents for Expression, Purification, and Labeling of Clathrin Light Chain

1. Construct/bacterial cells: pET28 LCA D203E C218S/*E. coli* BL21 DE3 [8].

2. LB medium containing kanamycin (30 μg/mL).

3. 1 M IPTG.

4. Lysis buffer 2: 20 mM BisTris pH 6.0 (*see* **Note 1**), 0.5 mM DTT, 1 mM EDTA, Complete protease inhibitor (*see* **Note 2**).

5. Mono Q 5/50 GL column, 1 mL bed volume (GE Life Sciences).

6. Buffer C (filter before use): 20 mM BisTris pH 6.0, 0.5 mM DTT.

7. Buffer D (filter before use): 20 mM BisTris pH 6.0, 1 M NaCl, 0.5 mM DTT.

8. Dialysis tubing or capsules (10 kDa MWCO).

9. Labeling buffer: 20 mM Tris–HCl pH 7.4, 1 mM EDTA, 0.2 mM tris(2-carboxyethyl)phosphine (TCEP).

10. Thiol-reactive dye, e.g., Alexa Fluor® 488 C5 Maleimide.

11. 2-Mercaptoethanol.

12. 0.5 M DTT.

13. Centrifugal filtration device (MWCO 10 kDa).

14. Glycerol.

2.4 Reagents for Expression and Purification of △PTEN-Auxilin

1. Construct/bacterial cells: pGEX auxilin 547-910 (N-terminal GST tag, thrombin cleavage site)/*E. coli* BL21 DE3 [11].

2. LB medium containing ampicillin (0.1 mg/mL).

3. 1 M isopropyl β-D-1-thiogalactopyranoside (IPTG).

4. Glutathione agarose.

5. Lysis buffer 3: 20 mM HEPES pH 7.6, 100 mM KCl, 0.2 mM EDTA, 20% glycerol, 1% Triton, 0.5 mM DTT, Complete protease inhibitor (Roche Life Science).

6. Wash buffer 1: 25 mM Tris–HCl pH 7, 40 mM NaCl, 0.1 mM EDTA, 0.2 mM TCEP.

7. Elution Buffer 1: 20 mM glutathione in wash buffer; make fresh; adjust pH to 8 with NaOH.

8. 0.5 U/μL thrombin, frozen in 50 μL aliquots.

9. Buffer E (filter before use): 50 mM MES, pH 6.7, 1 mM EDTA, 0.2 mM Tris(2-carboxyethyl)phosphine hydrochloride (TCEP).

10. Buffer F (filter before use): 50 mM MES, pH 6.7, 500 mM NaCl, 1 mM EDTA, 0.2 mM TCEP.

11. Mono S HR 5/5 column, 1 mL bed volume (GE Life Sciences).

2.5 Reagents for Expression, Purification, and Labeling of Hsc70

1. Construct/bacterial cells: pProEX Hsc70 (N-terminal hexahistidine tag, TEV cleavage site, C-terminal Gly-Cys extension)/*E. coli* BL21 DE3.

2. LB medium containing 0.1 mg/mL ampicillin.

3. 1 M IPTG.

4. Lysis buffer 4: 50 mM Tris–HCl pH 7.5, 300 mM NaCl, complete protease inhibitor tablet.

5. Cobalt affinity resin (e.g., Talon; (Clontech), or similar).

6. Wash buffer 2: 50 mM Tris–HCl pH 7.5, 300 mM NaCl, 10 mM imidazole.

7. Elution buffer 2: 50 mM Tris–HCl pH 7.5, 300 mM NaCl, 80 mM imidazole.

8. 0.5 M EDTA solution, pH 8.

9. 0.5 M DTT.

10. TEV protease.

11. 0.5 M Tris(2-carboxyethyl) phosphine (TCEP) stock solution.

12. 100 mM ATP.

13. Centrifugal filtration device, 30 kDa MWCO.

14. Superose 6 HR 10/30 column.

15. Column buffer 2: 20 mM imidazole pH 6.8, 100 mM KCl, 2 mM $MgCl_2$, 0.1 mM ATP.

2.6 Reagents for Purification of AP-2 Adaptor Protein

1. 4 fresh calf brains.

2. 50 mM phenylmethane sulfonyl fluoride (PMSF) solution in ethanol.

3. Buffer G: 50 mM MES, pH 6.9, 100 mM NaCl, 1 mM EGTA, 0.5 mM $MgCl_2$, 0.02% NaN_3, 0.5 mM DTT.

4. Buffer H (filter before use): 2.4 M Tris–HCl, pH 7.4, 0.04% NaN_3, 1 mM DTT.

5. 12.5% Ficoll/12.5% sucrose in Buffer G.

6. Column Buffer 3: 0.5 M Tris–HCl, pH 7.4, 0.02% NaN_3, 0.5 mM DTT.

7. Sepharose CL-4B column (length = 1 m, diameter = 3 cm, volume = 700 mL).

8. 0.5 M Ethylene glycol-bis(2-aminoethylether)-N,N,N',N'-tetraacetic acid (EGTA) pH 8.

9. Low phosphate buffer: 10 mM KH_2PO_4; pH 7.1, 100 mM KCl, 0.02% NaN_3, 0.1% 2-mercaptoethanol.

10. High phosphate buffer (*see* **Note 3**): 500 mM KH_2PO_4; pH 7.1, 100 mM KCl, 0.02% NaN_3, 0.1% 2-mercaptoethanol.

11. Ceramic hydroxyapatite column (2 mL bed volume, Type I (*see* **Note 4**), 20 microns).

12. AP buffer: 100 mM MES, pH 7.0, 150 mM NaCl, 1 mM EDTA, 0.02% NaN_3, 0.5 mM DTT.

2.7 Reagents for In Vitro Reconstitution of Fluorescent Clathrin Coat Formation

1. Recombinant clathrin heavy chain trimers.

2. Labeled recombinant clathrin LCA.

3. AP-2 adaptor complex purified from bovine brain.

4. Coat Formation Buffer (50 mM MES pH 6.5, 100 mM NaCl, 2 mM EDTA, 0.5 mM DTT) and/or Coated-Vesicle

Formation Buffer (80 mM MES pH 6.5, 20 mM NaCl, 2 mM EDTA, 0.4 mM DTT).

5. Dialysis cassette, 10 kDa MWCO.

2.8 Reagents for Fabrication of Liposomes Used for Reconstitution of Clathrin-Coated Vesicles

1. Lipids:
   (a) Maleimide-1,2-dioleoyl-sn-glycero-3-phosphoethanolamine (DOPE).
   (b) 1,2-dioleoyl-*sn*-glycero-3-phosphocholine (DOPC).
   (c) 18:1 PI(3)P 1,2-dioleoyl-*sn*-glycero-3-phospho-(1′-myo-inositol-3′-phosphate) (ammonium salt).
   (d) 18:1 PI(4)P 1,2-dioleoyl-sn-glycero-3-phospho-(1′-myo-inositol-4′-phosphate) (ammonium salt).
   (e) PI(4,5)P2   L-α-phosphatidylinositol-4,5-bisphosphate (Brain, Porcine) (ammonium salt).

2. Lipid Dyes:
   (a) DiD' oil; DiIC18(5) oil (1,1′-dioctadecyl-3,3,3′,3′-tetramethylindodicarbocyanine perchlorate)
   (b) DiI Stain (1,1′-dioctadecyl-3,3,3′,3′-tetramethylindocarbocyanine perchlorate ('DiI'; DiIC18(3))).

3. YQRL containing peptide (CKVTRRPKASDYQRLNL), lyophilized powder.

4. 20 mM HEPES pH 7.4.

5. Dimethyl sulfoxide (DMSO).

6. 10 mM β-mercaptoethanol.

7. Chloroform, methanol, and water (4:3:2.25 v/v mixture).

8. Avanti Lipid film extruder set: mini-extruder, two syringes (250 μL), and 50 nm pore size polycarbonate membrane.

2.9 Reagents for Assembly of Coats onto Liposomes

1. Coated-vesicle Formation Buffer: (80 mM MES pH 6.5, 20 mM NaCl, 2 mM EDTA, 0.4 mM DTT).

2. Extruded liposomes that are 50–80 nm in diameter (determined by negative stain EM).

3. Recombinant clathrin heavy chain triskelia.

4. Fluorescently labeled recombinant clathrin LCA.

5. AP-2 adaptor complex isolated from bovine brain (*see* **step 3**).

6. Dialysis cassette, 10 kDa MWCO.

2.10 Reagents for Fabrication and Assembly of Microfluidic Flow Cells

1. Silicon Wafer: 3″ N/Ph Orient. [111], 380umP, Mech Grade for Spin Coating or similar.

2. MicroChem SU-8 2050; Permanent epoxy negative photoresist.

3. MicroChem SU-8 developer.

4. TFOCS (Tridecafluoro-1,1,2,2, tetrahydrooctyl-1-trichloro silane).

5. Methanol.

6. Acetone.

7. Isopropanol.

8. Polydimethylsiloxane (PDMS): 184 sylgard elastomer and curing agent.

2.11 Reagents for Surface Chemistry on Glass Coverslips

1. Glass coverslips (#1.5).

2. Toluene.

3. Dichloromethane.

4. Ethanol.

5. 1:1 v/v ethanol–37% hydrochloric acid.

6. 1 M NaOH.

7. Poly-L-lysine(20 kDa) conjugated with PEG(2 kDa) and PEG-Biotin (3.4 kDa) (SuSoS AG, Switzerland).

2.12 Reagents for Capture of Clathrin Coats on the Surface of Coverslips

1. 1 mg/mL streptavidin in PBS.

2. Blocking solution: 20 mM Tris–HCl pH 7.5, 2 mM EDTA, 50 mM NaCl.

3. Biotinylated CVC.6 clathrin light chain A mouse monoclonal antibody.

2.13 Reagents and Microscope for Single-Particle Fluorescence Imaging Uncoating Assay

1. Uncoating Buffer:20 mM imidazole pH 6.8, 100 mM KCl, 2 mM $MgCl_2$, 5 mM protocatechuic acid, 50 nM protocatechuate-3,4-dioxygenase, 2 mM Trolox, 8 mM 4-nitrobenzyl alcohol.

2. Uncoating Mix: 1 μM Hsc70, 5 mM ATP, 10 mM $MgCl_2$, 25 nM ΔPTEN-auxilin1 in Uncoating Buffer.

3. Inverted fluorescence microscope equipped with the following hardware (Fig. 4).

   (a) Objective: 100×, 1.46NA TIRF oil objective, with an additional 2× magnification lens (spatial sampling necessary for point source model fitting for data collected using 16 μm² camera pixel size).

   (b) Stage controller (optional): Piezo 'Z' PZ-2000, X&Y Stage MS-2000 (Applied Scientific Instrumentation or similar).

   (c) TIRF slider unit: (Carl Zeiss Microimaging, Inc. or similar).

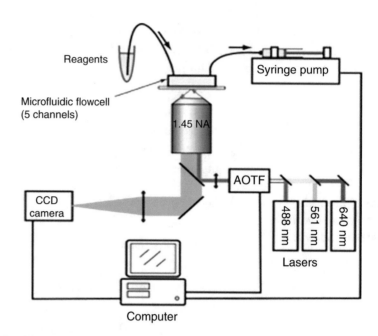

Reagents

Syringe pump

Microfluidic flowcell
(5 channels)

1.45 NA

AOTF

CCD
camera

488 nm
561 nm
640 nm

Lasers

Computer

Fig. 4 Microscope configuration. Schematic representation of the setup used for the single particle uncoating assay including the components for TIRF microscopy and the microfluidic flow cell

(d) Lasers: (λ = 488, 561, 647; 40–50 mW coupled with fiber optics to TIRF slider).

(e) Camera: EMCCD (Photometrics, Andor or similar), 512^2 imaging array (16 µm^2) creating an image pixel size of 0.08 µm.

3   Methodology

3.1   Expression and Purification of Clathrin Heavy Chain

3.1.1 Expression in Insect Cells

1. This procedure uses the Bac-to-Bac system (Invitrogen/ThermoFisher) to recombinantly express rat clathrin heavy chain based on a pFastBac-based plasmid encoding the heavy chain [10].

2. Transform DH10Bac cells with the clathrin heavy chain plasmid, and grow the cells on plates containing the appropriate antibiotics (gentamycin, tetracycline, kanamycin), Bluo-Gal, and IPTG. Completing blue–white screening, after 48 h, restreak a white colony. From this second plate, use a single colony to grow a 4 mL culture in LB medium containing the appropriate antibiotics (gentamycin, tetracycline, kanamycin). Isolate the bacmid DNA sample: lyse the cells with alkaline buffer, add 800 µL isopropanol, incubate on ice for 10 min., centrifuge at 18k × *g*, and wash the pellet with 70% ethanol.

Finally, air-dry the pellet in a fume hood, and resuspend it with Tris–EDTA buffer.

3. Prepare Sf9 cells using Sf900-II medium. Transfect cells in 6-well plates with the clathrin heavy chain bacmid according to the manufacturer's protocol.

4. After 3–5 days, isolate the cell supernatant to obtain a baculovirus stock.

5. Amplify the virus stock by infecting Sf9 cells and isolating the cell supernatant after considerable cell death is observed by Trypan Blue staining. Complete a second round of amplification.

6. Express clathrin heavy chain by using High Five cells grown in spinner flasks and Express Five or Excell 420 medium.

7. Prepare the cells to a density of $1.2–1.5 \times 10^6$ cells/mL (typically 2×500 mL), and infect them with the baculovirus stock (typically using an amount of the stock equal to 1% of the culture volume *see* **Note 5**).

8. Incubate in spinner flask for 2.5–3 days @ 28 °C.

9. Spin cells at 1500 rpm for 10 min in conical plastic bottles.

3.1.2 Cell Lysis and Purification of Clathrin Heavy Chain

10. Add one Complete tablet to 25 mL ice-cold Lysis Buffer 1. Resuspend the cells by quickly vortexing with 10 mL of the buffer, and then add remainder of the buffer.

11. Sonicate the material on ice for five periods of 60″ (with pauses of 120″) in a metal beaker.

12. Ultracentrifuge the lysate at 4 °C (e.g., TLX 100.4 rotor for 20 min at $438,813.5 \times g$).

13. Dialyze the supernatant in Cage Formation Buffer o/n at 4 °C; exchange once.

14. Spin the sample for 10 min at 1000 rpm in a table-top centrifuge; withdraw most of the supernatant (except for 2–3 mL), vortex, and spin again at 3000 rpm to recover more supernatant.

15. Ultracentrifuge the supernatant at 4 °C (e.g., Type 60 Ti for 1 h at $203,347.1 \times g$).

16. Supplement Buffer A with 2 mM PMSF or 0.6 mg/mL PEFA, and 3 mM EDTA. Using this solution, resuspend the pellet from **step 9** initially by pipetting, and then using a tissue grinder (Potter homogenizer; 5 strokes).

17. Add Buffer B (half the volume of that for Buffer A used in **step 10**), mix, and shake gently at room temperature for 15 min on a rocker.

18. Ultracentrifuge the sample in a TLA 100.4 for 20 min at 541,700 × g and 4 °C.

19. Filter the supernatant (0.2 μ syringe filter), and load onto a gel filtration (Sephacryl S-500) column equilibrated with Column Buffer.

3.1.3 Polishing Purification Step by Hydroxyapatite Chromatography

1. Use an FPLC at room temperature.

2. Wash the column with high phosphate buffer, and equilibrate it with low phosphate buffer.

3. Inject the sample onto the column, and wash by using five column volumes of low phosphate buffer.

4. Add 15 μL100 mM EGTA solution to the receiving tubes.

5. Elute the sample using a gradient of 0–100% high phosphate buffer.

6. Pool fractions that contain clathrin according to SDS-PAGE, and add glycerol to 20%. Aliquot the sample into 500 μL portions, and freeze by submerging the aliquots in liquid nitrogen for storage at −80 °C. Approximate expected concentration is 1 mg/mL.

3.2 Expression, Purification, and Labeling of Clathrin Light Chain

3.2.1 Expression in Bacteria

1. Inoculate 100 mL LB medium containing kanamycin (30 μg/mL) with a colony of BL21(DE3) bacteria transformed with a pET28-based vector for expression of rat light chain a1 (LCA) with mutations D203E and C218S (*see* **Note 6**). Grow an overnight culture at 37 °C with shaking at 250 rpm.

2. Inoculate 1 L LB medium containing kanamycin (30 μg/mL) in baffled flasks with 10 mL of saturated culture. Grow at 37 °C with shaking at 250 rpm until the OD(600) reaches approximately 0.5 (2.5–3 h).

3. Induce expression by addition of 0.6 mL 1 M IPTG (final concentration of 0.6 mM), and continue growth at 37 °C and 250 rpm for 3 h.

4. Pellet cells by centrifugation (5000 rpm, 4500 × g, 10 min, 4 °C).

5. Resuspend the bacterial cell pellet in cold Lysis Buffer 2 (volume ~20 mL).

6. Place suspension into conical flask and immerse in boiling water for 6 min, then chill on ice. The solution becomes turbid. Add fresh DTT solution to a concentration of 0.5 mM.

7. Centrifuge to pellet denatured protein and cell debris using a 60Ti rotor, 292,819.8 × g, 30 min, 4 °C. The supernatant contains the light chain.

8. Dialyze the supernatant against 1 L Lysis Buffer 2 (without protease inhibitors) at 4 °C overnight.

3.2.2 Purification of LCA Using Anion Exchange Chromatography

9. Wash MonoQ column with Buffer D, and the column with Buffer C.

10. Load filtered supernatant (0.2 μm syringe filter) into Superloop and inject onto column at a flow rate of 1 mL/min.

11. Elute the labeled light chain using a linear gradient from 0% to 32% Buffer D over 20 column volumes with a flow rate of 1 mL/min.

12. Add EDTA to each fraction to a final concentration of 1 mM.

13. Identify fractions containing LCA by SDS-PAGE. Concentrate combine fractions by centrifugal ultrafiltration. Approximate concentration of purified protein should be 1.5 mg/mL.

3.2.3 Labeling of LCA

14. Dialyze the protein solution against Labeling Buffer.

15. Determine the concentration of the protein from its absorbance at 280 nm (with the molar extinction coefficient of 32,430). The concentration in the labeling reaction should be between 3–6 mg/mL.

16. Add a 10 mM solution of the maleimide-functionalized fluorophore (*see* **Note** 7) to obtain a molar excess between 4:1 and 8:1. Mix and allow to react in the dark at room temperature for 2 h.

17. Quench the unreacted dye by addition of 2-mercaptoethanol to a final concentration of 10 mM.

18. Dilute the solution tenfold with Buffer C (without DTT) and remove the excess dye by centrifugal ultrafiltration. Repeat this process until the flow-through is colorless.

19. The labeled protein can be further purified with the chromatography conditions as described above in **step 11** using buffers without DTT and a linear gradient from 0–65% Buffer D over 40 column volumes.

20. Measure the UV-visible absorption spectrum of the sample. Determine the protein concentration and degree of labeling by using the extinction coefficient of the protein at 280 nm calculated from its sequence, the extinction coefficient of the fluorophore as provided by the manufacturer, and the correction factor for absorption of the fluorophore at 280 nm. Add glycerol up to 20% final concentration to the protein solution, divide into aliquots, and freeze in liquid nitrogen for storage at −80 °C.

3.3 Expression and Purification of ΔPTEN-Auxilin

3.3.1 Expression in Bacteria

1. Inoculate 100 mL LB medium containing ampicillin (0.1 mg/mL) with a colony of BL21(DE3) bacteria transformed with a pGEX vector for expression of auxilin 547-910. Grow an overnight culture at 37 °C with shaking at 250 rpm.

2. Inoculate 1 L LB medium containing ampicillin (1 mg/mL) in baffled flasks with 10 mL o/n culture and grow at 37 °C with

shaking at 250 rpm until the OD(600) reaches approximately 0.6–1.

3. Induce by addition of 0.25 mL 1 M IPTG (final concentration of 0.25 mM). Expression at 37 °C, 250 rpm, 4 h.

4. Pellet cells by centrifugation (5000 rpm, 4500 × g, 10 min, 4 °C).

5. Resuspend in 15 mL Lysis Buffer 3.

6. Lyse cells by sonication on ice.

7. Centrifuge the lysate for 30 min at 30,900 × g in a JA-17 rotor at 4 °C.

3.3.2 Purification of GST-Tagged Auxilin

8. Place 2 mL of glutathione–agarose slurry into a 15 mL conical tube, pellet beads (1500 rpm, 500 × g, 5 min), resuspend in 10 mL Lysis Buffer 3 to wash, pellet, decant supernatant. Repeat washing step. Resuspend beads in 0.5 mL Lysis Buffer 3.

9. Transfer beads into the lysate. Rotate end-over-end for 2 h.

10. Pour into column, let drain, collect flow-through.

11. Wash with 5–10 bed volumes of Lysis Buffer 3.

12. Wash with 5–10 bed volumes of Wash Buffer 1, let drain.

13. Place cap on column outlet; add 0.5 mL Elution Buffer 1, incubate for 5 min at RT, collect eluted proteins.

14. Repeat elution step five times.

15. Bradford assay or SDS PAGE to identify fractions with protein.

3.3.3 Removal of the GST Tag

16. Test digestion: (a) 20 μL GST-aux + 0.5 μL thrombin, (b) 20 μL GST-aux + 0.5 μL 1:4 dilution of thrombin in Elution Buffer 1. Incubate at room temperature and withdraw 10 μL samples after 1 h and 2 h. Boil samples with 10 μL Laemmli buffer and analyze on SDS PAGE with Coomassie staining.

17. Digestion: Scale up according to the results of the test digestion. Quench reaction by addition of complete protease inhibitor.

18. Add equal volume of Buffer E, adjust pH to ~6.6 (pH paper) by addition of 1 N HCl.

3.3.4 Separation of Auxilin from GST via Ion Exchange Chromatography

19. Spin protein solution (10 min, 18,000 × g) to remove aggregated material.

20. Inject protein solution onto Mono S column equilibrated with Buffer E.

21. Elute by running a linear gradient from 0% Buffer F to 80% Buffer F over 40 column volumes at a flow rate of 1 mL/min

(GST expected in the flow through; auxilin appears early in the gradient).

22. SDS PAGE to identify fractions containing auxilin, combine fractions, concentrate if needed (centrifugal filter, 10 kD MWCO).

23. Add 20% glycerol to the protein solution, divide into aliquots and freeze in liquid nitrogen for storage at −80 °C. Approximate protein yield should be 0.1–0.5 mg/mL.

3.4 Expression, Purification and Labeling of Hsc70

3.4.1 Expression in Bacteria

1. Inoculate 100 mL LB medium containing ampicillin (0.1 mg/mL) with a colony of BL21(DE3) bacteria transformed with a vector for expression of Hsc70 containing a Gly-Cys extension at the C-terminus and an N-terminal, TEV cleavable hexahistidine-tag. Grow an overnight culture at 37 °C with shaking at 250 rpm.

2. Inoculate 1 L LB medium containing ampicillin (0.1 mg/mL) in baffled flasks with 10 mL o/n culture and grow at 37 °C with shaking at 250 rpm until the OD(600) reaches approximately 0.3.

3. Transfer the cultures to 16 °C and allow to grow until the OD(600) reaches 0.4.

4. Induce by addition of IPTG (final concentration of 0.1 mM). Expression at 16 °C, 250 rpm, overnight.

5. Pellet cells by centrifugation (5000 rpm, 4500 × g, 10 min, 4 °C).

6. Resuspend the bacterial cell pellet in cold Lysis Buffer 4 (volume ~25 mL).

7. Break cells using a microfluidizer or by sonication.

8. Remove cell debris by centrifugation (45 Ti, 185,511.4 × g, 4 °C, 45 min).

3.4.2 Purification of Hexahistidine-Tagged Hsc70 Using Immobilized Metal Affinity Chromatography

9. Bind His-tagged protein in the soluble fraction (high speed supernatant) to Cobalt affinity resin (2 mL settled bed volume, equilibrated with Lysis Buffer 4) by head over head rotation at 4 °C for 2 h.

10. Pour beads into column, let drain.

11. Washing steps: (a) Lysis Buffer 4 (without protease inhibitor, 10 mL), (b) Wash Buffer 2 (10 mL).

12. Elute with Elution Buffer 2. Collect 0.5 mL fractions in tubes containing EDTA solution (final concentration of 2 mM).

13. SDS PAGE to identify fractions containing Hsc70.

14. Add DTT to a final concentration of 2 mM. Add TEV protease to a final concentration of 0.05 mg/mL.

15. Digest for 5 h at room temperature.

16. Dialyze against Lysis Buffer 4 (without protease inhibitor).

17. Incubate protein with Cobalt affinity resin (1 mL settled bed volume, equilibrated with Lysis Buffer 4) by head-over-head rotation at 4 °C for 2 h. TEV protease and Hsc70 with uncleaved hexahistidine tag binds to the beads during this step.

18. Pour beads into column, let drain and collect the flow-through (containing Hsc70 without hexahistidine tag).

19. Add 0.1 mM TCEP, 1 mM $MgCl_2$, and 0.1 mM ATP (final concentrations) and concentrate using a centrifugal filtration device (30 kDa MWCO). Centrifuge to remove aggregates (benchtop centrifuge, 20,000 × g, 4 °C, 10 min). Expected concentration ~2 mg/mL. Samples can be stored at −80 °C.

20. Add a thiol-reactive fluorophore (e.g., Alexa Fluor-C5-maleimide) in a ratio of 5:1 (fluorophore–protein, mol/mol) to the Hsc70 solution and incubate for 80 min at room temperature.

21. Quench unreacted fluorophore with 2-mercaptoethanol (final concentration of ~10 mM).

22. Separate the labeled Hsc70 from the free fluorophore by gel filtration on a Superose 6 HR 10/30 column operated at 0.5 mL/min with column buffer.

23. Identify and combine fractions containing labeled Hsc70 and concentrate by centrifugal ultrafiltration as described above in **step 19**.

24. Determine the protein concentration and degree of labeling using UV-visible absorption spectroscopy and the formula provided by the manufacturer. Add 20% glycerol to the protein solution, divide into aliquots and freeze in liquid nitrogen for storage at −80 °C.

3.5   Isolation of AP-2 Adaptor Protein (See Note 8)

3.5.1   Day 1 (4 °C Room):
Isolation of CCV from Brain
and Dissociation
of the Clathrin Coat at High
Tris Concentration

1. On the day prior to preparation: Store Buffer G, centrifuge bottles, and blender at 4 °C. Prepare Buffer B and store at room temperature. On the morning of preparation: Precool centrifuges to 4 °C. Prepare Ficoll/sucrose solution. Prepare Column Buffer 3 and store at room temperature.

2. Process brains immediately upon delivery. Cut the brains in halves, and wash them in a 4 L beaker containing cold tap water and ice.

3. Add 400 mL of Buffer G to the blender.

4. Keeping brains in ice-cold water, remove white matter and large vessels, cut remainder into cubes (~2.5 cm in each dimension), and place in blender. It is not essential to get rid of all white matter; just cut off the brainstem.

5. Add 10 mL of 50 mM PMSF solution, blend on a low setting for 1 min, then on a high setting for another 2 min. The sample should look homogeneous with a pinkish color.

6. Spin in a JA-10 rotor (Beckman J2-HS centrifuge) for 30 min at $17,650 \times g$ and 4 °C.

7. Carefully pour off supernatants.

8. Add 60 mL of Buffer G to remaining pellets and shake vigorously to resuspend; pour suspended pellets into blender and blend on a low setting for 1 min.

9. Spin the resuspended pellets in JA-14 rotor in Beckman J2-HS centrifuge for 45 min at $30,000 \times g$ and 4 °C.

10. Load the supernatants into Type 45 Ti tubes (70 mL).

11. Spin in Type 45 Ti rotor (Beckman) for 60 min at $224,500 \times g$ and 4 °C.

12. Following each spin, pour off and discard the supernatant. Resuspend 5 to 6 pellets by adding 10 mL Buffer G to only one tube, and then swirl to dislodge pellet. Pour this into next tube and remove that pellet. Repeat until all pellets are in solution. Add Buffer G until the total volume is 20 mL. Add PMSF from 50 mM stock (1:100, 200 μL).

13. Homogenize pellets with 5–10 strokes using a motor driven Potter homogenizer; make sure that no particulate is left.

14. Collect homogenate in a 50 mL conical tube. Wash pestle and glass homogenizer with Buffer G. Add this to homogenate and, if necessary, add more Buffer G until total volume is 25 mL. Add PMSF (1:200).

15. Add 25 mL of 12.5% Ficoll/12.5% sucrose to the homogenate. Mix well by repeated inversion.

16. Repeat **steps 15–18** with remaining pellets.

17. Pour these mixtures into four 25 mL ultracentrifuge tubes.

18. Spin in Type 60 rotor (Beckman) for 25 min at $90,400 \times g$ and 4 °C.

19. Collect supernatants.

20. Divide into four Type 45 Ti tubes (i.e., 70 mL tubes), fill to ~15 mL/supernatant per tube and fill them to the top with Buffer G. Mix by repeated inversion.

21. Spin in a Type 45 Ti rotor (Beckman) for 1 h. at $224,500 \times g$ and 4 °C.

22. Discard supernatants. Add 1 mL of Buffer G to only one tube, resuspend pellet, transfer suspension into a second tube and use it to resuspend the pellet in the second tube. Transfer into homogenizer. Use another 1 mL of Buffer G to rinse both tubes. Repeat procedure with remaining tubes until all pellets are resuspended. Total volume must not exceed 10 mL.

23. Homogenize pellets (coated vesicles) as in **step 16**.

24. Pour into a 50 mL conical tube. Add Buffer G until volume is 10 mL, if necessary.

25. Add 5 mL of Buffer H and 300 μL of 50 mM PMSF to initiate depolymerization of coats.

26. Mix well by repeated inversion and leave overnight at RT.

3.5.2 Day 2 (Room Temperature): Gel Filtration to Separate Clathrin from Adaptor Proteins (APs)

27. Put 15 mL sample into a Type 60 Ti tube (25 mL tube volume). Fill to the top with water to reduce Tris concentration to approx. 0.4 M.

28. Spin for 1 h at $203,300 \times g$ and room temperature in Type 60 Ti rotor (Beckman).

29. Carefully remove and save the supernatant except for ~2 mL on top of the fluffy pellet. Collect this remaining volume, spin it again in the TLA-100.4 rotor (TLX centrifuge; Beckman) for 20 min at $541,700 \times g$ and room temperature, and add it to the rest. Pass the solution through a syringe filter (0.2 μm) and store it in a 50 mL conical tube.

30. Load the sample onto a Sepharose CL-4B column and run at 1 mL/min. Collect fractions of 5 mL. Clathrin appears as a broad peak after the void, followed by a small peak containing AP-1 and AP-2.

31. Pool clathrin fractions, add EGTA to a final concentration of 1 mM and store at 4 °C. Pool AP fractions and store at 4 °C overnight.

3.5.3 Day 3 (4 °C Room): Hydroxyapatite Chromatography to Separate AP-1 and AP-2

32. Connect the hydroxyapatite column to the FPLC system at 4 °C. Put a 0.2 μm syringe filter at the inlet of the column. Operate at 1 mL/min for all steps. Wash the column with 20 mL of high phosphate buffer, and equilibrate it with low phosphate buffer.

33. During the equilibration phase, load the 50 mL Superloop with the AP sample.

34. Inject the APs, and continue running low phosphate buffer.

35. Elute AP-1 and AP-2 from the column using the following gradient program: 0–60% high phosphate buffer over 24 mL, and then a step to 100% high phosphate buffer, with continued elution for 20 mL (*see* **Note 9**).

36. Pool AP-1 fractions and AP-2 fractions separately, and dialyze them against 1 L of AP buffer overnight, and for a few more hours after exchanging the buffer (4 °C).

37. Add glycerol to 20%, and freeze the samples for storage at −80 °C.

3.6 In Vitro Reconstitution of Fluorescent Clathrin and Fluorescent Coat Formation

1. Mix a solution of recombinant clathrin heavy chain with a solution of labeled clathrin LCA at a molar ratio of heavy chain–LCA = 1:2.4 and incubate at room temperature for 40 min. Total protein concentration should be 1 mg/mL.

2. Add a solution of AP-2 to the clathrin solution prepared in **step 1** to a ratio of clathrin–AP-2 of 3:1 (w/w). Mix the solution and transfer to a dialysis capsule (*see* **Note 10**).

3. Dialyze overnight against Coat Formation Buffer at 4 °C. Replace Coat Formation Buffer and dialyze for an additional 3–4 h.

4. Transfer the coat solution to a 1.5 mL centrifuge tube, centrifuge to remove larger aggregates (benchtop centrifuge, 15,000 RCF, 4 °C, 10 min).

5. Transfer supernatant to fresh centrifuge tube, centrifuge to collect coats (TLA-100.4, 228,887.3 × g, 4 °C, 12–16 min).

6. Immediately withdraw supernatant with a 1 mL pipette. The pellet should have a hemispherical dome shape and be colorless and translucent.

7. Wash carefully with Coat Formation Buffer around the pellet.

8. Add Clathrin Coat Formation Buffer or Coated-Vesicle Formation Buffer to the tube (use a volume of 120 μL for a pellet with a diameter of ~3 mm), allow to stand at room temperature for 10–15 min in the dark, then slowly wash buffer over the pellet using a micropipettor to resuspend, avoiding foaming.

9. Coats can be stored frozen at −80 °C in the presence of 20% glycerol (snap-freeze in liquid nitrogen).

3.7 Liposome Preparation

3.7.1 YQRL-Peptidolipid Synthesis

10. Prepare a mix of 20 mg/mL of YQRL containing peptide (CKVTRRPKASDYQRLNL; prepared in 20 mM HEPES buffer pH 7.4), DMSO, and maleimide-DOPE (1:1:2 v/v mixture respectively).

11. Vortex the mixture at 1000 rpm for 2 h at room temperature.

12. Add 10 mM β-mercaptoethanol to the coupling reaction in order to quench any unreacted groups.

13. Extract the YQRL peptidolipid by adding chloroform, methanol, and water (4:3:2.25 v/v mixture) and centrifuging at 1000 rpm for 5 min.

14. Dry the organic phase containing the peptidolipid under argon and stored in a sealed argon atmosphere containing vial.

15. Resuspend the films in chloroform and methanol mixture (2:1) at 2 mg/mL (based on the mass in the sealed lipids vials) prior to use for liposome lipid film prep.

3.7.2 Phosphoinositide Specific Lipid Films

16. Mix the desired lipid composition (default Molar percent: 2% PI(4,5)P$_2$, 5% PI(4)P or PI(3)P, 3% YQRL-DOPE peptidolipid, 0.1% DiI or DiD lipid dye, 86.9% DOPC) in 20:9:1 chloroform–methanol–water

17. Dry the mixture under argon to prepare composition specific lipid films and store as 300 μmol aliquots at −20 °C in argon purged sealed chambers until the expiration of the lipids as specified by the manufacturer.

3.7.3 Lipid Film Extrusion

18. Immediately prior to reconstitution of the clathrin-coated vesicles, hydrate the lipid film aliquots by adding 300 μL of Coat Formation Buffer and letting sit at room temperature for 3 min.

19. Vortex the mixture using a benchtop mixer at max speed for 30 s. Let sit for 60 s and repeat this step.

20. Assemble the mini-extruder (membrane supports and 50 nm pore size polycarbonate membrane), and inject 200 μL of Coat Formation Buffer through the extruder.

21. Aspirate the solution into the extrusion syringe and inject through the extruder ~20–30 passes to generate liposomes with a 50–80 nm diameter.

3.8 Reconstitution of Clathrin-Coated Vesicles (CCV)

1. Clathrin-coats composed of clathrin and AP-2 self-assemble around liposomes (containing YQRL-peptidolipid and PI(4,5) P$_2$.)

3.8.1 Day 1

2. Premix clathrin heavy chain and fluorescently labeled light chain (1:3 mol/mol ratio) at room temperature for 20 min. Minimum volume required is 300 μL.

3. Add AP2 (3:1 w/w clathrin:AP2) and extruded liposomes (15 μL of extruded liposomes per 100 μg of Clathrin heavy chain, with 2% PI(4,5)P$_2$, 5% PI(4)P or PI(3)P, 3% YQRL-DOPE peptidolipid, 0.1% DiI or DiD lipid dye, 86.9% DOPC) to the clathrin heavy chain and light chain mixture.

4. Using a 200 μL pipette to gently mix the solution without introducing bubbles or foam.

5. Transfer the above solution into a mini dialysis cassette (10 kDa MWCO) and dialyze at 4 °C overnight against Coated-vesicle Formation Buffer.

6. Replace the dialysis buffer and redialyze for an additional 4 h at 4 °C.

7. Transfer the dialyzed CCV mixture into an 1.5 mL tube and spin down at max speed on a benchtop centrifuge at 4 °C for 10 min in order to remove large aggregates.

8. Transfer the supernatant into a fresh tube and centrifuged in a TLA-100.4 at $265,500 \times g$ for 30 min at 4 °C.

9. Withdraw and discard the supernatant.

10. Using a micropipettor resuspend the CCV containing pellet in Coated-vesicle Formation Buffer (in 100 μL of buffer per 100 μg of clathrin heavy chain; avoid foaming) and centrifuge a second time at $265,500 \times g$ for 30 min at 4 °C.

11. Withdraw and discard the supernatant.

12. Resuspend the final pellet in Coated-vesicle Formation Buffer and use the CCVs immediately for best results.

3.9   Fabrication and Assembly of Microfluidic Flow Cells in a Clean Room Facility

3.9.1   Channel Design

1. To prevent channels from collapsing or sagging, the height (h) and width (w) of channels, and space (s) between channels should correspond to the following optimized values: $h = 50$ μm (variable and controlled below using the spin speed and viscosity of the photoresist); $w = 200$ μm; $s = 500$ μm.

2. Design the mask (using CAD program of your choice) while optimizing the number of channels per chip, and number of chips per silicon wafer (default: 5–7 channels per chip, 10–13 chips per wafer, typical channel length 10 mm).

3.9.2   Fabrication and Photolithography: (3–5 Wafers Prepared in Parallel in a Clean Room)

3. Pour negative photoresist on silicon wafer, and spin-coat for 50 μm (or desired) thickness.

4. Place mask over the silicon wafer with prebaked (on a hot plate) photoresist and expose to UV (duration based on thickness of the photoresist).

5. Develop (to remove non-UV exposed photoresist) wafer to create a master wafer, using developing times specified by the manufacturer.

3.9.3   Passivate the Master Wafer: (Ensure That the Surface Is Nonsticky)

6. Place wafer in vacuum desiccator with a petri dish containing TFOCS (tridecafluoro-1,1,2,2, tetrahydrooctyl-1-trichlorosilane) for 30 min to deposit a monolayer.

7. This step is critical to make sure that PDMS does not stick to the master wafer.

8. Repeat the passivation process after casting >20–30 PDMS molds.

3.9.4   PDMS Molding

9. Pour and mix PDMS and curing agent in 10:1 ratio and degas to remove bubbles.

10. Cure the PDMS mold for at least 6 h at 70 °C.

11. Carefully peel the cured PDMS mold from the master wafer, and cover with adhesive Scotch tape.

12. Using micropuncher, punch holes at the channel edges to create an inlet and outlet (*see* **Note 11**).

3.9.5 Device Assembly

13. Cut and shape the PDMS device with inlet and outlet punched holes to fit the imaging coverslip.

14. Plasma treat (by glow discharging) precleaned coverslip and PDMS for 20–40 s at 0.35 mBar pO_2 (using purged oxygen plasma; air plasma may also be used, however, bonding conditions should be optimized since the quality of adhesion varies based on atmospheric compositions, humidity levels, power, etc.; *see* **Note 12**) [12].

15. Immediately join the two plasma exposed surfaces, press gently to remove any air bubbles and heat for 5 min at 100 °C on a hot plate to improve bonding strength.

16. Functionalize devices immediately using PLL-PEG solution, or store devices in clean room until ready for use.

3.10 Surface Chemistry on Glass Coverslips

1. Sequentially rinse the glass coverslips for 20 min in the following solvents in series: toluene, dichloromethane, ethanol, ethanol–hydrochloric acid (1:1 v/v), and Milli-Q water. Store coverslips in water until ready to assemble the PDMS device (*see* Subheading 3.9.5 above).

2. The freshly bonded chips are functionalized by the following series of channel infusions (injected volumes for channel washes should be at least 100× the channel volume):

   (a) Flushing 1 mL of 1 M NaOH at 200 μL/min.

   (b) Flushing 2 mL of ddH_2O at 200 μL/min.

   (c) Immediately followed by PLL-PEG/PLL-PEG-Biotin solution 20 μL at 10 μL/min and let sit for 5 min.

   (d) Wash free reagents from the channels by flowing 100 μL of ddH_2O at 20 μL/min.

   (e) Functionalize the PLL-PEG-Biotin groups by flowing in 20 μL of streptavidin in Blocking Solution at 5 μL/min and let sit for 5 min (20 μL of 1 mg/mL streptavidin in PBS + 80 μL 20 mM Tris–HCl pH 7.5, 2 mM EDTA, 50 mM NaCl).

   (f) Wash excess Streptavidin by flowing in 100 μL of ddH_2O at 20 μL/min.

   (g) Functionalize the inert layer of PLL–PEG–Biotin–Streptavidin in the channels with biotin-labeled CVC.6 clathrin light chain A mouse monoclonal antibody by flowing in 20 μL of 10 μg/mL at 10 μL/min and let sit for 5 min.

3.11 Single-Particle Uncoating Assay

*3.11.1 Microscope Calibration (See **Note 13**)*

TIRF Angle

1. If the TIRF angle of the microscope needs to be calibrated for quantitative imaging, prepare a coverslip with immobilized fluorescent beads (100–170 nm diameter).

2. Add ~200 μL of MilliQ water containing fluorescent beads, and place on the microscope in order to calibrate the TIRF angle.

3. Set the TIRF slider angle such that the laser illumination is passing straight through the objective (*warning*: laser illumination will no longer be confined within the microscope).

4. Set the laser power and exposure such that the emission counts from the immobilized beads are ~2–5% of the EMCCD's max dynamic range.

5. At this point, you will observe a mix of immobilized beads and free-floating beads.

6. Optimize the evanescent filed: Adjust the TIRF slider angle to maximize the signal from the immobilized beads and minimize signal from the floating beads. Ideally, set the angle between 70 and 80% of maximum TIRF, as this will reduce the background fluorescence from injected solutions.

Single-Molecule Calibration

7. Prepare a coverslip with immobilized fluorophores in the femtomolar to picomolar-range required to resolve single fluorophores (labeled clathrin LCA adsorbs to glass coverslips cleaned as in Subheading 3.10, **step 1** followed by glow discharge; oxygen plasma is not required; however, conditions should be optimized since the extent of adsorption varies depending on atmospheric compositions, humidity levels, etc.).

8. Adjust the power and exposure such that you are able to detect the signal from single fluorophores with a signal-to-noise ratio of 5 or higher. Fix the power and image and subsequently bleach the fluorophores; repeat this process at different exposures and extract the single bleaching step-sizes in order to generate a single molecule fluorescence signal calibration curve.

9. Adjust the laser power such that imaging is carried out with one fluorophore sensitivity with exposures of 150–200 ms.

3.11.2 Capture of Clathrin Coats or Reconstituted Coated Vesicles on the Surface of Functionalized Coverslips

1. Inject 15 μL of Clathrin Coats (CC) or Clathrin Coated Vesicles (CCVs) (in Coat Formation Buffer or Clathrin Coated Vesicle Formation Buffer respectively) at 10 μL/min into the CVC.6 antibody functionalized channel and let sit for 1–5 min (the concentration of CC or CCV may be titrated if the field-of-view is saturated, or conversely too few CC or CCVs are bound).

2. Wash the free CC or CCVs using the Uncoating Buffer.

3. Subsequently, inject the Uncoating Mix at 20 µL/min for 150 s into the channel to disassemble the CCs or CCVs.

3.11.3 Fluorescence Microscopy of Uncoating:

1. Prior to the uncoating assay:

   (a) Calculate the injection tubing volume and the time needed for injected uncoating reagent mix to enter the microfluidic channel by injecting a testing fluorescent solution (such as fluorescein).

   (b) Calibrate the TIRF angle (*see* above subheading "TIRF-Angle calibration") of the microscope and ensure the system is configured between 70% and maximum TIRF angle.

2. Uncoating Assay:

   (a) Start microscope acquisition at 1 Hz using 50 ms exposure for each of the excited channels 10 s prior to when the uncoating mixture enters into the channel and end acquisition after 150 s.

   (b) Monitor the signal of fluorescent clathrin excited using the 488 nm laser.

   (c) If required, monitor the signal of fluorescent Hsc70 generated by exciting it with the appropriate laser.

   (d) If required, monitor the signal of the fluorescent lipid dye in liposomes using the appropriate laser.

   (e) Quantify signals from CCs, or CCVs using a fixed position 2D point source detector [13].

   (f) Using single-molecule intensity calibration (*see* Subheading 3.11, **steps 7–9**) for the fluorescent clathrin light chain A, convert the recorded signal amplitude for the clathrin coats to the number of molecules.

4 Notes

1. Adjust the pH of the BisTris buffer at room temperature. The pH of the solution at 4 °C is approximately pH 6.5.

2. Add DTT and Complete protease inhibitor from stock solutions (stored at −20 °C) just before using the buffer.

3. The potassium salt is used because it has higher solubility at 4 °C than the sodium salt.

4. The separation of AP-2 from AP-180 is improved with type I ceramic hydroxyapatite compared to type II.

5. We determine the amount of viral stock to use for batch production based on an experiment completed with High Five

cells in well-plates: wells are treated with a series of volumes of virus stock, and protein production is assessed by SDS-PAGE and Coomassie staining.

6. The cDNA of rat clathrin light chain a1 was cloned into the pET28 expression vector using the NcoI and EcoRI restriction sites to allow expression without tags. The point mutations were introduced using site-directed mutagenesis. The mutation D203E restores the epitope recognized by the monoclonal antibody CVC.6 (which is used to capture clathrin coats onto the glass coverslip surface for imaging). The mutation C218S removes one of the two cysteines in the light-chain sequence to allow site-specific labeling at C187 with thiol-reactive reagents.

7. The dye can be stored in aliquots at −20 °C for an extended period of time (>1 year). Prepare a 10 mM stock solution of the dye in water, divide into aliquots (e.g., 10 μL in 0.2 mL microcentrifuge tubes) and evaporate to dryness for storage in an airtight container containing a desiccant. The dye aliquots can be redissolved in water or buffer and added to the desired concentration to the protein solution for labeling.

8. Biochemistry 23, 4420–4426 (1984).

9. The elution profiles for AP-1 and AP-2 tend to vary considerably from one purification to the other; AP-1 is eluted first and tends to be eluted from the column in three to four 1 mL fractions. The step to 100% buffer B in the gradient is meant to coincide with the elution of AP-2 and was included to elute AP-2 at high concentration from the column. The fractions containing the APs need to be verified by SDS-PAGE.

10. The volume of the solution for coat formation is typically 0.5 mL. The concentration of clathrin heavy chain should be around 1 mg/mL.

11. Use only nitrile gloves, not latex, when handling PDMS to preserve adhesive functionality.

12. Over exposure to plasma may damage surfaces and affect bonding strength.

13. *Caution*: Please make sure that appropriate laser safety precautions are observed during the microscope calibration and imaging experiments.

Acknowledgments

T.B. was supported by grants from the Australian Research Council (DP130100936, FT 1001004) and the National Health and Medical Research Council (1098870, 1100771).

The Kirchhausen laboratory was generously supported by grants from Biogen and by National Institutes of Health Grants R01 GM075252.

References

1. Ehrlich M, Boll W, Van Oijen A, Hariharan R, Chandran K, Nibert ML, Kirchhausen T (2004) Endocytosis by random initiation and stabilization of clathrin-coated pits. Cell 118:591–605

2. Lee DW, Wu X, Eisenberg E, Greene LE (2006) Recruitment dynamics of GAK and auxilin to clathrin-coated pits during endocytosis. J Cell Sci 119:3502–3512

3. Massol RH, Boll W, Griffin AM, Kirchhausen T (2006) A burst of auxilin recruitment determines the onset of clathrin-coated vesicle uncoating. Proc Natl Acad Sci U S A 103: 10265–10270

4. Merrifield CJ, Perrais D, Zenisek D (2005) Coupling between clathrin-coated-pit invagination, cortactin recruitment, and membrane scission observed in live cells. Cell 121: 593–606

5. Barouch W, Prasad K, Greene L, Eisenberg E (1997) Auxilin-induced interaction of the molecular chaperone Hsc70 with clathrin baskets. Biochemistry 36:4303–4308

6. Braell WA, Schlossman DM, Schmid SL, Rothman JE (1984) Dissociation of clathrin coats coupled to the hydrolysis of ATP: role of an uncoating ATPase. J Cell Biol 99:734–741

7. Ungewickell E, Ungewickell H, Holstein SE, Lindner R, Prasad K, Barouch W, Martin B, Greene LE, Eisenberg E (1995) Role of auxilin in uncoating clathrin-coated vesicles. Nature 378:632–635

8. Bocking T, Aguet F, Harrison SC, Kirchhausen T (2011) Single-molecule analysis of a molecular disassemblase reveals the mechanism of Hsc70-driven clathrin uncoating. Nat Struct Mol Biol 18:295–301

9. Fotin A, Cheng Y, Grigorieff N, Walz T, Harrison SC, Kirchhausen T (2004) Structure of an auxilin-bound clathrin coat and its implications for the mechanism of uncoating. Nature 432:649–653

10. Rapoport I, Boll W, Yu A, Bocking T, Kirchhausen T (2008) A motif in the clathrin heavy chain required for the Hsc70/auxilin uncoating reaction. Mol Biol Cell 19:405–413

11. Fotin A, Cheng Y, Sliz P, Grigorieff N, Harrison SC, Kirchhausen T, Walz T (2004) Molecular model for a complete clathrin lattice from electron cryomicroscopy. Nature 432:573–579

12. Millare B, Thomas M, Ferreira A, Xu H, Holesinger M, Vullev VI (2008) Dependence of the quality of adhesion between poly(dimethylsiloxane) and glass surfaces on the conditions of treatment with oxygen plasma. Langmuir 24:13218–13224

13. Aguet F, Antonescu CN, Mettlen M, Schmid SL, Danuser G (2013) Advances in analysis of low signal-to-noise images link dynamin and AP2 to the functions of an endocytic checkpoint. Dev Cell 26:279–291

Chapter 11

Spatial and Temporal Aspects of Phosphoinositides in Endocytosis Studied in the Isolated Plasma Membranes

Ira Milosevic

Abstract

Endocytosis is a well-orchestrated cascade of lipid–protein and protein–protein interactions resulting in formation and internalization of vesicles. Membrane phospholipids have key regulatory functions in endocytosis and membrane traffic. I have previously described an in vitro assay based on the isolated, substrate-attached plasma membrane to study the spatial distribution and levels of phosphoinositides, in particular phosphatidylinositol-4,5-bisphospate [$PI(4,5)P_2$]. This assay utilizes cultured cells subjected to a brief ultrasonic pulse, resulting in the formation of thin, flat inside-out plasma membrane sheets with elements of cell cytoskeleton. Here, I describe an experimental procedure for "on-stage" and "off-stage" detection of $PI(4,5)P_2$ spatial distribution, and semi-quantification of $PI(4,5)P_2$ levels in the plasma membrane using fluorescence microscopy. Depending on the probe selected for lipid detection, this simple assay can be modified to study other plasmalemmal phospholipids and/or proteins.

Key words Endocytosis, Phospholipids, Plasma membrane, Cell-free assay, Isolated membrane sheets, Phosphoinositides, $PI(4,5)P_2$, Lipid probe

1 Introduction

Membrane lipids provide a structural scaffold for the organization of cellular enzymatic machineries and for the anchoring of the cytoskeleton. They are also key participants in the regulation of cellular metabolism and functions due to their fast turnover. Endocytosis is a fast-paced process by which cells internalize membrane proteins, lipids and soluble molecules. The best characterized internalization pathway is clathrin-mediated endocytosis, however, several faster clathrin-independent pathways are presently intensely researched [1–5]. Those include macropinocytosis, ultrafast endocytosis, kiss-and-run, activity-dependent bulk endocytosis, and fast-endophilin-mediated endocytosis. Since these endocytic processes occur on a millisecond to second timescale, they demand rapid cellular

The original version of this chapter was revised. A correction to this chapter can be found at
https://doi.org/10.1007/978-1-4939-8719-1_19

Laura E. Swan (ed.), *Clathrin-Mediated Endocytosis: Methods and Protocols*, Methods in Molecular Biology, vol. 1847,
https://doi.org/10.1007/978-1-4939-8719-1_11, © Springer Science+Business Media, LLC, part of Springer Nature 2018

reaction to avert exhaustion or overstimulation. Membrane phosphoinositides and their metabolism pose key regulatory elements in a number of endocytic processes and membrane traffic in general, and their importance is slowly starting to be appreciated [6–9].

Phosphoinositides can accomplish various cellular functions in part due to their low abundance and rapid and controlled turnover [3, 7, 10, 11]. As regulators of endocytosis, they have several advantages over proteins with the same task. Phosphoinositides can recruit and activate a large number of various proteins to the plasma membrane to create a local environment where endocytic processes could take place ([6, 9]). Yet, at the same time, the rapid enzymatic production and degradation of phosphoinositides allow the cell to remain flexible: by changing the phosphoinositide-level cells can change the distribution of entire sets of proteins within milliseconds to seconds, thereby modifying physiological function without the need for protein synthesis or degradation. Recent advancements in imaging techniques (e.g., superresolution) and development of novel, more specific probes for phosphoinositides detection facilitated a number of spatial and temporal studies that have contributed a better understanding of these unique molecules (reviewed in [8]). In addition, by combining techniques for detection and manipulation of phosphoinositides, many specific roles of individual phosphoinositides are recently unveiled [9].

Notably, several systems are amenable to address questions regarding the temporal and spatial interrelation of lipids and proteins involved in endocytosis [8, 9]. I focus here on a simple and well-defined cell-free assay that was originally developed by Avery and collaborators (2000). This assay utilizes attached cultured cells that are sheared by a short ultrasonic pulse, resulting in the formation of a several nanometer thin, flat inside-out membrane sheet (Figs. 1, 2, 3). This assay has subsequently been combined with a number of probes and methods to visualize various phospholipids and proteins [12–17]. Major advantages of this technique are its simplicity, affordability and the feature that the membrane and membrane-associated structures can be distinguished by fluorescence microscopy without any interference of labeled structures above or below the focal plane, resulting in a high signal to noise ratio. Basically, the resolution of lipids and proteins in the plane of the membrane is limited only by the resolution of the objective lens of the light microscope.

In comparison to methods with an analogous spatial resolution (e.g., total internal reflection fluorescence (TIRF) microscopy), isolated plasma membrane sheets allow direct and controllable biochemical access to the plasma membrane [13, 14, 18]. For example, the endocytic (and also exocytic) machinery can be directly studied using membrane-impermeable biochemical tools like fluorescently labelled recombinant proteins or antibodies. Though the use of this in vitro assay is not universal and other assays might be more suitable depending on the nature of studies, this technique has been and is being used to unravel some of the complex reac-

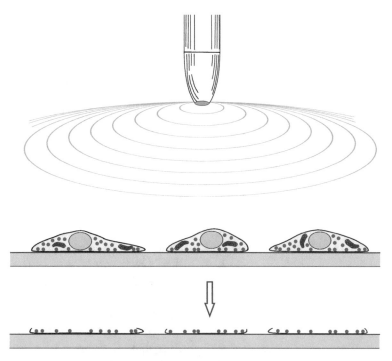

Fig. 1 Schematic of isolated plasma membrane sheet preparation by sonication

Fig. 2 Example of the experimental setup for studying isolated membranes "on-stage." (**a**) Photograph of the microscopic setup used for "on-stage" sonication of cultured chromaffin or clonal cells. (**b**) Detailed view of a sonication chamber made of plexiglas and containing a glass coverslip with cells on the bottom. The titanium sonication tip points into the chamber's lumen. (**c**) Schematic of isolated plasma membrane sheet preparation by sonication "on-stage"

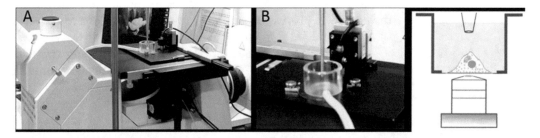

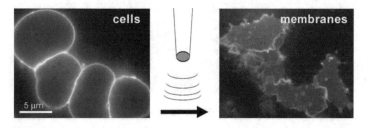

Fig. 3 Plasma membrane sheets from bovine chromaffin cells in primary culture generated "on-stage" in the field of view of a fluorescent microscope. The membranes can be viewed are using 1-(4-trimethylamoniumphenyl)-6-phenyl-1,3,5-hexatriene (TMA-DPH), a lipophilic styryl dye that intercalates between phospholipids and it is visible in the blue channel

tions leading to fission and the fate of molecules in budding vesicles prior and after the fission [17, 19].

We have used plasma membrane sheets to detect the levels and spatial organization of numerous plasmalemmal proteins as well as phospholipid phosphatidylinositol-4,5-bisphosphate [PI(4,5)P$_2$], the best characterized phosphoinositides (Fig. 4; [14]). We employed the pleckstrin homology (PH) domain of phospholipase Cδ_1 fused to enhanced green fluorescent protein (EGFP) to visualize PI(4,5)P$_2$ in the plasma membranes of primary cells (chromaffin and neuronal cells), and a number of cultured clonal cell lines (PC12, HeLa, HEK293, mouse embryonic fibroblasts). Using the membrane sheet assay, we observed that PI(4,5)P$_2$ was concentrated in discrete microdomains on the plasma membrane (Fig. 4; [14]). Overexpression of PI(4,5)P$_2$-generating enzyme, phosphatidylinositol-4-phosphate-5-kinase Iγ (PI4P5KIγ), or the brief (up to 5 min) application of the inhibitor of phosphatidylinositol 3-kinase LY294002, led to a rapid increase in the plasmalemmal PI(4,5)P$_2$ levels [14]. Expression of a membrane-targeted inositol 5-phosphatase domain of synaptojanin 1, or prolonged (over 30 min) application of LY294002 have strongly reduced PI(4,5)P$_2$ levels in the plasma membrane. Both PI(4,5)P$_2$ levels and its spatial distribution in the plasma membrane are essential for the endocytic process, as shown by numerous studies and reviewed by Saheki and De Camilli [3] and Posor et al. [9].

The detailed experimental procedure for detecting transient PI(4,5)P$_2$ levels and its spatial distribution at the plasma membrane using fluorescence microscopy is described below.

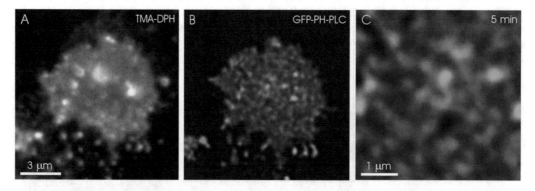

Fig. 4 PI(4,5)P$_2$ microdomains in the isolated plasma membrane of bovine chromaffin cells. (**a**) Freshly prepared membrane sheet from chromaffin cell stained with a lipid dye TMA-DPH and visualized in the blue channel. (**b**) Freshly isolated plasma membrane sheet is incubated for 5 min with 3 µM EGFP-PH-PLCδ_1, washed and fixed in 4% PFA for 15 minutes. The observed punctuate staining suggests an existence of PI(4,5)P$_2$-microdomains in the plasma membrane of chromaffin cells. (**c**) Magnified view from B: EGFP-PH-PLCδ_1 binding to the plasma membrane

2 Materials

2.1 Buffers

1. Sonication buffer with 300 nM $[Ca^{2+}]_{free}$: 120 mM potassium glutamate, 20 mM potassium acetate, 20 mM Hepes, 2 mM ATP, 100 μM GTP, 4 mM $MgCl_2$, 4 mM EGTA, 6 mM Ca^{2+}-EGTA, pH 7.2, $[Ca^{2+}]_{free}$ = 300 nM. Osmolarity should be ~310 mOsm/kg. After preparation, the buffer is filtered through 0.22 μm filter, and pH was adjusted to 7.2–7.3. Sonication buffer is aliquoted in 50 ml tubes and kept at −20 °C for up to 3 months. The defrosted sonication buffer is kept on ice, and bubbled with N_2 for 30 min. Before being used in experiments, 2 mM ATP and 100 μM GTP are added.

2. Locke's buffer: 154 mM NaCl, 5.6 mM KCl, 0.85 mM NaH_2PO_4, 2.15 mM Na_2HPO_4, 10 mM glucose, pH 7.4, osmolarity 305-310 mOsm/kg.

3. Substrate coating buffer: poly-L-lysine in 100 mM sodium borate, pH 8.5.

4. Phosphate buffered saline (PBS): 137 mM NaCl, 2.7 mM KCl, 10 mM Na_2HPO_4, 1.76 mM KH_2PO_4, pH 7.3-7.4; sterilize by autoclaving.

5. Blocking-buffer: PBS with 3% BSA.

6. K-Glu buffer: 120 mM potassium glutamate, 20 mM potassium acetate, 20 mM HEPES, and 4 mM $MgCl_2$, 4 mM EGTA, 6 mM Ca^{2+}-EGTA, pH 7.2, $[Ca^{2+}]_{free}$ = 300 nM.

7. 100 mM ethylene glycol-O,O′-bis(2-aminoethyl)-N,N,N',N'-tetraacetic acid.

8. (EGTA) in deionized water. *See* Subheading 3.1 for preparation. Store at −20 °C.

9. 100 mM Ca-EGTA solution (*see* details in Subheading 3.1): 1.0001 g $CaCO_3$, 3.804 g EGTA add 50 ml ultrapure water. Add 20 ml of freshly diluted 1 M KOH, and 0.238 g HEPES, heat to 70C to dissolve $CaCO_3$. Cool, add KOH until pH 7.2 at room temperature. Bring volume to 100 ml with ultrapure water. Store at −20 °C.

2.2 Reagents and Consumables

1. 37% hydrochloric acid (HCl).

2. 1 N KOH.

3. 0.1 mg/ml poly-L-lysine. Store at −20 °C.

4. TMA-DPH   (1-(4-trimethylammoniumphenyl)-6-phenyl-1,3,5-hexatriene *p*-toluenesulfonate) dye.

5. 2 mM ATP in deionized water. Store at −20 °C.

6. 100 μM GTP in deionized water. Store at −20 °C.

7. 6- and 12-well sterile tissue culture dishes.

8. Syringe filter unit 0.22 μm for sterilization of solutions that cannot be autoclaved.

9. 1 mg/ml collagenase in Locke's buffer.

10. 10 ml syringe.

11. Nylon mesh (50 μm pore size).

12. *Escherichia coli* BL21 (DE3).

13. Luria Broth: 20 g tryptone, 20 g NaCl, 10 g yeast extract, adjust pH to 7.0 using 5N NaOH. Bring to 2 l with deionized water and autoclave.

14. 50 mg/ml kanamycin (1000× stock). Store at −20 °C.

15. 500 mM isopropyl-1-thio-D-galactopyranoside (1000× stock).

16. pET28a-His6-EGFP-PH-PLCδ$_1$ plasmid and mutant containing the mutations K30A, K32A and W36N.

17. 3-[(3-Cholamidopropyl)dimethylammonio]-1-propanesulfonate (CHAPS).

18. 200 mM phenylmethane sulfonyl fluoride (PMSF) solution in ethanol.

19. Lysozyme.

20. Triton X-100.

21. 2 M imidazole.

22. Thrombin (approximately 500 units/mg protein).

23. SDS-PAGE gel, and electrophoresis apparatus.

24. Coomassie blue protein stain.

25. Nickel–nitrilotriacetic acid (Ni-NTA) agarose.

26. Bradford assay kit.

27. 100% ethanol.

2.3 Cell Cultures and Media

1. PC12 cells (PC-12 ATCC® CRL-1721™ *Rattus norvegicus* adrenal gland pheochromocytoma).

2. Chromaffin cell growing medium: DMEM with 4.5 g/l glucose supplemented with 10 ml/l serum substitute containing insulin, transferrin and selenium (e.g., ITSX; Gibco/Life Technologies), penicillin, and streptomycin.

3. PC12 growth medium: DMEM with 4.5 g/l glucose supplemented with 10% heat-inactivated fetal calf serum (FCS), 5% horse serum, penicillin, and streptomycin.

4. Trypsin–EDTA (0.05%).

5. Bovine adrenal glands (to be obtained fresh from slaughterhouse).

2.4 Equipment

1. Ca^{2+}-selective electrode for measuring free calcium concentration in buffers and solutions (We use Orion™ Calcium Electrodes model 9720BNWP Ionplus, ThermoFisher Scientific).

2. Large sonicator (e.g., Sonifier 450, Branson Ultrasonics Corp., Danbury, CT) with a small probe diameter (tip ø2.5 mm) is used to disrupt the cultured cells without removing the basal plasma membranes. The power setting is 1.8, and a duty cycle is 100 ms.

3. Small portable sonicator (e.g., Sonifier B-12, Branson Ultrasonics Corp. Danbury, CT) with a small probe diameter (tip ø1 mm) is used to disrupt the cultured cells and generate isolated membrane sheets "on-stage" and while imaging. The ultrasonic pulse with defined power setting (output control 5.4; 500 ms duration) is applied.

4. Zeiss Axiovert 100 TV fluorescence microscope (other microscope brands/models can be used) with a 100× 1.4 NA plan achromate objective, fluorescence filters compatible with TMA-DPH (blue) dye; and EGFP is used for imaging of samples. Throughout all experiments the focal position of the objective was controlled using a low voltage piezo translator driver and a linear variable transformer displacement sensor/controller The images were taken with a back-illuminated frame transfer CCD-camera (2 × 512 × 512-EEV chip, 13 × 13 μm pixel size, Princeton Instruments Inc.) with a magnifying lens (1.6× Optovar), and analyzed with Metamorph software (Molecular Devices).

5. Oven (to dry $CaCO_3$ and EGTA powders).

6. Benchtop shaker.

7. Bunsen burner.

3 Methods

3.1 Preparation and Purity Check of EGTA Stock Solutions

The calcium chelator ethylene glycol-O,O′-bis(2-aminoethyl)-N,N,N′,N′-tetraacetic acid (EGTA) is used for the fine adjustment of nanomolar and micromolar levels of free calcium ($[Ca^{2+}]_{free}$). 100 mM EGTA and 100 mM Ca-EGTA stock solutions are prepared first. Before weighing the chemicals, $CaCO_3$ is baked at 110 °C for 1 h, and EGTA at 150 °C for 3 h. pH is adjusted using 1 N KOH, the solutions were filtered and pH was adjusted to 7.2 (*see* **Note 1** for pH adjustment). The purity of EGTA stock solutions and the concentration of free Ca^{2+} ions in the solutions are checked using a Ca^{2+}-selective electrode (*see* **Note 2**).

3.2 Cleaning and Coating of Glass Coverslips with Poly-L-lysine

Glass coverslips (ø15 mm for chromaffin cells/ø25 mm for PC12 cells) are washed with 1 M HCl for 30 min at room temperature (using shaker), and rinsed 3 times with deionized water. Next, coverslips are rinsed 2 times with 100% ethanol and subsequently stored in ethanol until further use. Each coverslip is flamed for 1 s, and placed in a well in a 12-well dish. Coverslips are then treated with a 0.1 mg/ml poly-L-lysine solution for 30 min at 37 °C in a cell culture incubator. After coating, the poly-L-lysine solution is removed and coverslips are washed with sterile deionized water. If kept at 4 °C in a plastic bag, coated coverslips can be used for up to 1 week.

3.3 Cell Preparation (2–3 Days Before Experiment)

Primary cultures of bovine chromaffin cells are used for the experiments. Two or three bovine adrenal glands can be obtained from the local slaughterhouse and transported to the laboratory in Locke's buffer on ice. After the fat and connective tissue is trimmed, 10 ml of Locke's buffer is injected in each gland through the main vein (this step needs to be repeated 3 times). Subsequently, the glands are injected with 1 mg/ml collagenase in Locke's buffer, and incubated for 15 min at 37 °C. After collagenase treatment, the medulla is dissected out from each gland, put in 10 ml Locke's buffer and centrifuged at $190 \times g$ for 2 min at 20 °C. The supernatant is removed and the tissue is minced through a nylon mesh, while adding a total of 50 ml Locke's buffer. The cell suspension is subjected to 8 min centrifugation at $100 \times g$ and 20 °C: after each centrifugation step the pellet were resuspended in Locke's buffer, and this procedure is repeated three to four times until mainly chromaffin cells were present in the preparation (and red blood cells are washed away). The cells are resuspended in chromaffin cell growing medium (see **Note 3**). The cells are plated on ø15 mm coated glass coverslips (~5×10^5–7×10^5 cells/coverslip) and kept at 37 °C in 8% CO_2. They are used for experiments 1–3 days after plating.

If bovine adrenal glands cannot be obtained, it is possible to culture chromaffin cells also from mice, but such primary cultures are not pure (in addition to chromaffin cells, such primary cultures contain many cells that originate from the cortex of adrenal medulla; [20, 21]). In addition, cultured PC12 cells can also be used as a model system. PC12 are cultured in 75 cm² uncoated flasks in PC12 growth medium at 37 °C in 8% CO_2. Before splitting, cells are washed twice with PBS, detached with 4 ml trypsin–EDTA for 2 min, harvested and centrifuged at $235 \times g$ and 20 °C for 5 min. The pellet is resuspended in PC12 growth medium (~6×10^5 cells/ml). 0.5 ml aliquots were seeded on ø25 mm coverslips coated with poly-L-lysine (Subheading 3.2). After additional 15 min to allow the cells to settle, 2.5 ml of PC12 growth medium is added. Cells were used for the experiments about 24 h after.

3.4 Expression and Purification of Wild-Type and Mutated EGFP-PH-PLCδ1

Escherichia coli BL21 (DE3) cells are transformed with wild-type (wt) or mutated (K30A, K32A, W36N) pET28a-His6-EGFP-PH-PLCδ$_1$ and grown in 2 l Luria broth medium containing 50 μg/ml kanamycin at 37 °C until their optical density at 600 nm reached about 0.5–0.6. Protein expression is induced by the addition of 0.5 mM isopropyl-1-thio-D-galactopyranoside for 4 h. The bacteria are harvested by centrifugation. The pellet is resuspended in 35 ml of ice-cold PBS containing 1 mM PMSF, and incubated with lysozyme for 10 min on ice. After sonication, 1% Triton X-100 and 1% CHAPS are added. The mixture is incubated for another 10 min and then centrifuged at 25,000 × *g* for 15 min. The supernatant is supplemented to 20 mM imidazole and incubated with 4 ml Ni-NTA agarose for 2 h at 4 °C. The collected beads are washed twice with PBS buffer containing 20 mM imidazole, once with PBS buffer containing 40 mM imidazole, and once with K-Glu buffer. The beads are subsequently resuspended in 2 ml of K-Glu buffer supplemented with 10 mM EGTA, and after 30 min, thrombin (30 U) is added and incubated for 4 h. The supernatant containing His$_6$-tag-cleaved protein is dialyzed twice against 0.5 L K-Glu buffer supplemented with 1 mM dithiothreitol. Finally, the dialyzed solution is sedimented in a microcentrifuge (14,000 rpm for 30 min at 4 °C), and the protein concentration of the supernatant is determined by the method of Bradford. The protein is analyzed by SDS-PAGE and Coomassie staining. Two bands are detected: one of the appropriate size (75% of the total protein amount) and another significantly smaller, yet only the bigger band associates specifically with PI(4,5)P$_2$ liposomes.

3.5 Generation of Membrane Sheets from Cultured Chromaffin Cells or PC12 Cells "Off-Stage"

To generate plasma membrane sheets in this cell-free assay, cells can be sonicated "off-stage" or "on-stage" (*see* **Note 4**). For" off-stage" sonication, sonication buffer is defrosted, kept on ice and bubbled with N$_2$ for 30 min or until saturated with air. Chromaffin cells are cultured on poly-L-lysine coated coverslips for at least 24 h (maximum 72 h). Membrane sheets from cultured chromaffin and/or PC12 cells are generated as previously described [18]. Briefly, 150 ml of ice-cold sonication buffer (*see* **Note 5**) is placed in round glass beaker (ø9.5 cm) with a final volume of 300 ml. Coverslips with cells are placed in the beaker and cells are sheared using a single ultrasound pulse in the sonication buffer. For sonication, a tip diameter of 2.5 mm, a coverslip-to-tip clearance of 12 mm and a duty cycle of 100 ms are used as standard, with power setting as a variable (usually 1.8). Immediately after sonication, coverslips with isolated membrane sheets are incubated with 3 μM EGFP-PH-PLCδ$_1$ in K-Glu buffer with 3% BSA for 1–3 min at cold plate (4 °C). Subsequently, the sheets are washed 3 times for 30 s in K-Glu buffer and fixed for 1 h in 4% paraformaldehyde in PBS at room temperature in the dark. Before imaging, the sample is washed with PBS, and incubated with 50 mM NH$_4$Cl in PBS for

15 min to eliminate free aldehyde groups. For visualizing the samples, a coverslip is moved to the imaging chamber and 50 μl of saturated   1-(4-trimethyl-amoniumphenyl)-6-phenyl-1,3,5-hexatriene (TMA-DPH) solution is added to 700 μl PBS. Alternatively, a specific marker that labels plasma membrane (e.g., mRFP-CAAX) can be expressed in cells prior to plasma membrane preparation. The imaging is performed as detailed below.

3.6 Generation of Isolated Membrane Sheets from Cultured Chromaffin and PC12 Cells "On-Stage"

In addition to the aforementioned "off-stage" approach, the membrane sheets can be also made "on-stage" in the field of view of a light microscope (*see* **Note 4**; Figs. 2 and 3). When chromaffin cells are sheared directly in the visual field of microscope ("on-stage sonication"), a specially designed round chamber (ø28 mm) with a final volume of 6 ml and a sonication tip (ø1 mm) with a coverslip-to-tip clearance of 8 mm are used. To visualize the cells, 50 μl saturated aqueous solution of TMA-DPH dye is added to 6 ml sonication buffer (Fig. 3). The ultrasonic pulse with defined power setting (output control 5.4; 500 ms duration) is applied and the membrane sheets are generated while imaging. The coverslip is exposed to a single pulse. However, since the area of membrane sheet formation is local under described conditions, up to 4 ultrasonic pulses could be applied upon moving the chamber in respect to sonication tip. Immediately upon sonication, the 5/6 of sonication buffer is removed through a suction system, and EGFP-PH-PLCδ₁ is added up to final concentration of 3 μM and incubated for 30s to 3 min, depending on desired intensity. The sample is imaged as detailed below during labeling and subsequent washing steps (2 times 5 ml K-Glu buffer).

3.7 Immunofluorescence on Chromaffin and PC12 Isolated Membrane Sheets

When isolated membrane sheets need to be fixed with paraformaldehyde immediately after being generated (e.g., for detection of plasma membrane proteins; *see* **Note 4**) the sonication was performed using a procedure similar to "off-stage" preparation (*see* Subheading 3.5). A tip with a diameter of 2.5 mm, a coverslip-to-tip clearance of 10 mm and a duty cycle of 100 ms are used. The standard power setting at Sonifier 450 is 1.8–2.0. A coverslip with attached cells is centered under the sonication tip in a beaker with 150 ml ice-cold sonication buffer, and a single pulse is applied (the debris resulting from cell sonification is negligible when diluted in 150 ml of sonification buffer).

Subsequently, the isolated sheets are fixed for 30 min at room temperature in 4% paraformaldehyde in PBS, and washed twice with PBS. The free aldehyde groups are blocked by incubation in PBS with 50 mM NH_4Cl for 15 min, and two washing steps with PBS were performed next. Immunocytochemical labeling is carried out using the indirect fluorescence method. In short, the membrane sheets were incubated with the blocking solution and then with a primary antibody for 1–3 h at room temperature, or overnight at 4 °C. The samples are then washed three times in PBS for 10 min each. An hour-long incubation with the secondary antibody is performed at room temperature and in the dark. Before imaging, the membrane sheets were washed three times in PBS for 10 min each, and subsequently mounted in the imaging chamber with 0.5 ml PBS.

3.8 Digital Imaging by Fluorescence Microscopy

The samples are examined with a Zeiss Axiovert fluorescence microscope with aforementioned specifics. Since it is important to keep the sample in focus, throughout all experiments the focal position of the objective was controlled using a low voltage piezo translator driver and a linear variable transformer displacement sensor/controller. Given that the sample is thin, and there is no interference of labelled structures above or below the focal plane, epifluorescent microscope is as good choice as more advance spinning-disk microscope or a TIRF microscopy setup.

For fixed samples, single images are taken with a video CCD-camera controlled by Metamorph software. To minimize photo bleaching, the exposure time of 500 ms was usually used. One image is always taken in the presence of TMA-DPH dye to visualize the plasma membrane.

For the on-stage imaging, the samples are imaged for 10–15 min in the K-Glu buffer. The image is taken every 5–30 s (depending on the experimental design) with a CCD-camera controlled by Metamorph software. Various reagents or dyes can be added during imaging (e.g., methyl-β-cyclodextrin that removes cholesterol from the membrane). The changes in the number or distribution of fluorescent dots during longer imaging intervals are quantified using the Cantata module of the Khoros image analysis software (Khoral Research, Albuquerque, USA).

3.9 Quantification of Fluorescent Signal

To perform comparative quantitation of fluorescence intensity, membrane sheets are identified in the TMA-DPH images. A 4.1×4.1 μm (50×50 pixels, 1 pixel corresponds to 81.25 nm) region of interest is defined on the membrane and then transferred to the other channels. Occasionally the correction of the region position is made to avoid obvious artifacts like highly fluorescent contaminating particles that are occasionally seen. The fluorescence intensity is quantified by measuring the average intensity of the area. The local background is measured in the area outside the membrane sheets and subtracted. For each condition, 60–120 membrane sheets are analyzed. Intensity values are given as mean ± S.E.M. As a spatial reference, TetraSpeck™ microfluorospheres of an average size of 0.22 μm (Molecular Probes, Eugene, Oregon, USA) can be detected in all channels.

4   Notes

1. Since two protons are released upon calcium binding to one EGTA molecule, the addition of $CaCO_3$ into 100 mM EGTA solution makes the solution acidic. Given that released protons react with the CO_3^{2-} ions and form carbonic acid, it is possible to reduce the acid content by driving CO_2 out of the solution. This can be done either various means, for example by bubbling the solution with N_2 for 30 min, or by heating it to 80 °C.

2. The concentration of free Ca^{2+} ions in the solution, $[Ca^{2+}]_{free}$, is measured by a Ca^{2+}-selective electrode (Orion). The Ca^{2+}-selective electrode is filled with 100 mM $CaCl_2$, and the silver wire is chlorided. The calcium concentration of a test solution is determined using one of two procedures: by calibration of electrode with a series of solutions of known concentrations, or by titration with $CaCl_2$ with concomitant pH control. The calibration of the Ca^{2+}-selective electrode is commonly performed with a series of following concentration: 1, 3, 10, 30, 100, and 300 μM, and 1, 3 and 10 mM $[Ca^{2+}]_{free}$ solutions. When the electrode is immersed in a solution of a known concentration, the potential is recorded by pH meter (on the mV scale). The potential changes by approximately 29 mV for a divalent ion for every decay of a charge in the ion concentration. The readings are used to make a calibration chart, where the x-axis represents the log of ion concentration, and the y-axis the mV recordings. The conditions during calibration of the Ca^{2+}-selective electrode are selected to match the measuring conditions of a test solution as closely as possible (volume, pH, temperature, osmolarity, ionic composition). Notably, the concentration of EGTA in the stock solution can also be checked by adding a slight excess of calcium and titrating pH with KOH. At pH 7.2, EGTA releases 1.978 protons per calcium ion [22].

3. The chromaffin cells are resuspended and maintained in the chromaffin cell growing medium, i.e., DMEM medium supplemented with insulin, transferrin and selenium. Supplementation of basal medium with insulin, transferrin and selenium standardizes the conditions under which the cells are cultured and eliminates the need for serum supplementation, i.e., addition of fetal bovine serum (FBS). The lack of FBS reduces the attachment of many adherent cell types: this is particularly useful to increase the ratio of chromaffin cells in the chromaffin cell primary cultures. Insulin is known to promote glucose and amino acid uptake, lipogenesis, monovalent cation and phosphate transport, protein and nucleic acid syntheses. Transferrin serves as a carrier for iron and may help to reduce

toxic levels of oxygen radicals. Selenium acts as an antioxidant in media and as a cofactor for glutathione peroxidase and other proteins.

4. Plasma membrane sheets can be obtained from cultured cells attached to the surface of the glass coverslip by sonicating the cells "on-stage" or "off-stage." For detection of acute events at the plasma membrane and/or detection of lipids in the plasma membrane, it is best to generate the plasma membrane sheets directly on the top of the objective (in the visual filed of the microscope) so the membrane sheets can be imaged directly and without any delay. Yet, if the protein abundance and distribution need to be studied, for convenience the isolated membrane sheets can simply be prepared "off-stage," and immediately fixed with paraformaldehyde at room temperature or precooled methanol at −20 °C. I have never noticed any difference in the plasma membrane sheet quality or characteristics that depended on the way the sheets are generated. Due to the nature of experiment, the "off-stage" approach yields many more sheets (usually several millions).

5. Concentration of $[Ca^{2+}]_{free}$ in the sonication buffer depends on the desired experimental conditions; I most commonly use 300 nM $[Ca^{2+}]_{free}$.

Acknowledgment

I thank Dr. A. Milosevic and Dr. M. Barszczewski for kind help with the figures, and Dr. N. Raimundo for comments. The author declares no competing financial interests. This work is supported by the grants of the German Research Foundation (DFG) through the collaborative research center SFB-889 (project A8) and SFB-1190 (project P02), and the Emmy Noether Young Investigator Award (1702/1).

References

1. Kononenko NL, Haucke V (2015) Molecular mechanisms of presynaptic membrane retrieval and synaptic vesicle reformation. Neuron 85(3):484–496

2. McMahon HT, Boucrot E (2012) Molecular mechanism and physiological functions of clathrin-mediated endocytosis. Nat Rev Mol Cell Biol 12(8):517–533

3. Saheki Y, De Camilli P (2012) Synaptic vesicle endocytosis. Cold Spring Harb Perspect Biol 4(9):a005645

4. Milosevic I. (2018) Revisiting the Role of Clathrin-Mediated Endoytosis in Synaptic Vesicle Recycling. Front Cell Neurosci.12:27. https://doi.org/10.3389/fncel.2018.00027.

5. Watanabe S, Boucrot E (2017) Fast and ultra-fast endocytosis. Curr Opin Cell Biol 47:64–71

6. Cremona O, De Camilli P (2001) Phosphoinositides in membrane traffic at the synapse. J Cell Sci 114(Pt 6):1041–1052

7. De Craene JO, Bertazzi DL, Bär S, Friant S (2017) Phosphoinositides, major actors in

membrane trafficking and lipid Signaling pathways. Int J Mol Sci 18(3):E634

8. Idevall-Hagren O, De Camilli P (2015) Detection and manipulation of phosphoinositides. Biochim Biophys Acta 1851(6):736–745

9. Posor Y, Eichhorn-Grünig M, Haucke V (2015) Phosphoinositides in endocytosis. Biochim Biophys Acta. 1851(6):794–804. https://doi.org/10.1016/j.bbalip.2014.09.014. Epub 2014 Sep 28. Review. PMID: 25264171

10. Vicinanza M, D'Angelo G, Di Campli A, De Matteis MA (2008) Function and dysfunction of the PI system in membrane trafficking. EMBO J 27(19):2457–2470

11. Wenk MR, De Camilli P (2004) Protein–lipid interactions and phosphoinositide metabolism in membrane traffic: insights from vesicle recycling in nerve terminals. Proc Natl Acad Sci U S A 101:8262–8269

12. de Wit H, Walter A, Milosevic I, Gulyás-Kovács A, Sørensen JB, Verhage M (2009) Four proteins that dock secretory vesicles to the target membrane. Cell 138(5):935–946

13. Lang T, Bruns D, Wenzel D, Riedel D, Holroyd P, Thiele C, Jahn R (2001) SNAREs are concentrated in cholesterol-dependent clusters that define docking and fusion sites for exocytosis. EMBO J 20:2202–2213

14. Milosevic I, Sørensen JB, Lang T, Krauss M, Nagy G, Haucke V, Jahn R, Neher E (2005) Plasmalemmal phosphatidylinositol-4,5-bisphosphate level regulates the releasable vesicle pool size in chromaffin cells. J Neurosci 25(10):2557–2565

15. Nagy G, Milosevic I, Fasshauer D, Müller M, de Groot B, Lang T, Wilson MC, Sørensen JB (2005) Alternative splicing of SNAP-25 regulates secretion through non-conservative substitutions in the SNARE domain. Mol Biol Cell 16:5675–5685

16. Nagy G, Milosevic I, Mohrmann R, Wiederhold K, Walter AM, Sørensen JB (2008) The SNAP-25 linker as an adaptation toward fast exocytosis. Mol Biol Cell 19(9):3769–3781

17. Wu M, Huang B, Graham M, Raimondi A, Heuser JE, Zhuang X, De Camilli P (2010) Coupling between clathrin-dependent endocytic budding and F-BAR-dependent tubulation in a cell-free system. Nat Cell Biol 12(9):902–908

18. Avery J, Ellis DJ, Lang T, Holroyd P, Riedel D, Henderson RM, Edwardson JM, Jahn R (2000) A cell-free system for regulated exocytosis in PC12 cells. J Cell Biol 148:317–324

19. Wu M, De Camilli P (2012) Supported native plasma membranes as platforms for the reconstitution and visualization of endocytic membrane budding. Methods Cell Biol 108:3–18

20. Sørensen JB, Nagy G, Varoqueaux F, Nehring RB, Brose N, Wilson MC, Neher E (2003) Differential control of the releasable vesicle pools by SNAP-25 splice variants and SNAP-23. Cell 114(1):75–86

21. Sørensen JB, Wiederhold K, Müller M, Milosevic I, Nagy G, de Groot B, Grubmüller H, Fasshauer D (2006) Sequential N- to C-terminal "zipping-up" of the SNARE complex drives priming and fusion of secretory vesicles. EMBO J 25(5):955–966

22. Smith GL, Miller DJ (1985) Potentiometric measurements of stoichiometric and apparent affinity constants of EGTA for protons and divalent ions including calcium. Biochim Biophys Acta 839(3):287–299

Chapter 12

SMrT Assay for Real-Time Visualization and Analysis of Clathrin Assembly Reactions

Devika Andhare, Sachin S. Holkar, and Thomas J. Pucadyil

Abstract

Clathrin-mediated endocytosis manages the vesicular transport of the bulk of membrane proteins from the plasma membrane and the *trans*-Golgi network. During this process, discrete sets of adaptor proteins recognize specific classes of membrane proteins, which recruit and assemble clathrin lattices on the membrane. An important determinant to the success of this vesicular transport reaction is the intrinsic ability of adaptors to polymerize clathrin on a membrane surface. Adaptor-induced clathrin assembly has traditionally been analyzed using static electron microscopy-based approaches. Here, we describe a methodology to follow adaptor-induced clathrin assembly in real-time using fluorescence microscopy on a facile model membrane assay system of *supported membrane tubes* (SMrT). Results from such assays can be conveniently run through routine image analysis procedures to extract kinetic parameters of the clathrin assembly reaction.

Key words Clathrin, Adaptors, Clathrin assembly, Endocytosis, Fluorescence microscopy, Model membranes

1 Introduction

Clathrin-mediated endocytosis (CME) is an essential cellular process that manages the vesicular transport of the bulk of membrane proteins from the plasma membrane and the *trans*-Golgi network [1–3]. The initial stages of CME involve recognition of specific sets of membrane proteins and lipids by a conserved family of adaptor proteins, which subsequently recruit and assemble clathrin lattices on the membrane. Recent genome-wide RNAi screens have revealed CME to be orchestrated by a large repertoire of adaptor and accessory proteins [4]. In vitro reconstitution approaches are therefore fundamental to arriving at a detailed mechanistic understanding of the contribution of each participant protein to the dynamics and extent of clathrin assembly.

The original version of this chapter was revised. A correction to this chapter can be found at
https://doi.org/10.1007/978-1-4939-8719-1_19

Laura E. Swan (ed.), *Clathrin-Mediated Endocytosis: Methods and Protocols*, Methods in Molecular Biology, vol. 1847,
https://doi.org/10.1007/978-1-4939-8719-1_12, © Springer Science+Business Media, LLC, part of Springer Nature 2018

Previous literature describes several end-point biochemical assays to analyze adaptor–clathrin interactions but necessitate electron microscopy to assess the success of a clathrin assembly reaction leaving the dynamics of this process essentially unclear [5–7]. Recent fluorescence microscopy-based analysis of adaptor-induced clathrin assembly on giant unilamellar vesicles (GUVs) offers the possibility of analyzing dynamics of clathrin assembly [8]. Nevertheless, the utility of GUV-based assays is limited since (a) vesicles are difficult to tether to the surface, thus disallowing flowing in clathrin to temporally control the assembly process, (b) such reactions, as is the case with previous EM-based analyses, inherently suffer from clathrin assembly in solution into empty cages since the excess unbound adaptors cannot be washed off before clathrin is introduced, and (c) assembled clathrin on a spherical membrane surface (such as on a vesicle) is subject to Brownian diffusion which necessitates complex particle tracking approaches for analysis of dynamics of clathrin assembly.

To circumvent all of the above limitations, we have recently developed a fluorescence microscopy-based assay to monitor adaptor-induced clathrin assembly on a system of supported membrane tubes (SMrT), a template that is essentially a diffraction-limited membrane tube noncovalently pinned to a passivated glass coverslip and assembled inside a small volume flow cell [9–11]. The primary motivation for the use of membrane tubes is based on our recent observation that adaptor-mediated clathrin assembly is facilitated on a highly curved membrane surface [12]. The membrane tubes are coated with a desired adaptor and excess unbound adaptor is washed off from solution thereby restricting clathrin assembly to the adaptor-coated membrane tube. The clathrin assembly reaction is then initiated by flowing in fluorescently labeled native clathrin and manifests in the formation of fluorescent punctae on the membrane tube. The entire process is restricted to a pseudo one-dimensional surface and therefore allows interrogation of reaction dynamics at the level of single events using standard image analyses of kymographs generated of the assembly reaction. In this chapter, we describe a detailed methodology for the analysis of clathrin assembly induced by epsin, a conserved monomeric adaptor for ubiquitinylated membrane proteins.

2   Materials

2.1   Recombinant and Native Protein Purification

1. LB broth.
2. BL21DE3 competent *E. coli*.
3. Ampicillin.
4. Autoinduction medium for bacterial growth (e.g., Formedium, UK).

5. HisPur cobalt resin.

6. Empty polypropylene columns for gravity-based flow (e.g., PD-10 columns (Amersham)).

7. StrepTRAP (5 ml) column (GE Healthcare).

8. Anion exchange resin (e.g., Q Sepharose (1 ml) column from GE Healthcare).

9. Waring blender.

10. Dounce homogenizer.

11. Peristaltic pump.

12. FPLC setup (e.g., AKTA Prime Plus chromatograph system from GE Healthcare).

13. HEPES buffered-saline (HBS): 20 mM HEPES, 150 mM NaCl, pH 7.4.

14. Phenylmethanesulfonyl fluoride (PMSF): Reconstituted as a 100 mM stock in isopropanol.

15. d-Desthiobiotin.

16. HisPur cobalt resin elution buffer: HBS containing 100 mM imidazole, pH 7.4.

17. StrepTRAP elution buffer: HBS containing 2.5 mM desthiobiotin, pH 7.4.

18. Cage assembly buffer: 100 mM MES, 1 mM EGTA, 0.5 mM $MgCl_2$, pH 6.8.

19. Floatation buffer: 12.5% w/v Ficoll 400 and 12.5% w/v sucrose solution prepared in cage assembly buffer.

20. Cage disassembly buffer: 10 mM Tris–HCl, pH 8.0.

21. Anion exchange elution buffer: 25 mM Tris–HCl, 1 M NaCl, pH 8.0.

22. Ammonium sulfate.

23. 10% SDS-PAGE components.

2.2 Fluorescent Labeling of Clathrin

1. Texas Red C2 maleimide dye.

2. DMSO.

3. 1 M DTT: 1 M stock prepared in water and stored at −20 °C.

2.3 Passivation of Glass Coverslips

1. 40 mm round glass coverslips.

2. 3N sodium hydroxide.

3. Piranha solution: Concentrated H_2SO_4 and 30% H_2O_2 mixed at a 3:2 v/v ratio.

4. Glycidyloxypropyltrimethoxysilane (GOPTS).

5. Acetone.

6. Polyethylene glycol 8000 (PEG8000).

7. Dry heating block capable of reaching 100 °C.

2.4 Preparing SMrT Templates

1. FCS2 closed chamber system equipped with a 0.1 mm thick 14 × 24 silicone gasket and a micro perfusion low flow pump (Bioptechs, PA).

2. 1,2-Dioleoyl-*sn*-glycero-3-phosphocholine (DOPC). Stocks reconstituted in chloroform to a final concentration of 25.4 mM.

3. 1,2-Dioleoyl-*sn*-glycero-3-phospho-L-serine (sodium salt) (DOPS). Stocks reconstituted in chloroform to a final concentration of 12.3 mM.

4. 1,2-Dioleoyl-*sn*-glycero-3-phospho-(1′-myo-inositol-4′,5′-bisphosphate) (ammonium salt) (DOPI(4,5)P2). Stocks are reconstituted in methanol to a final concentration of 0.91 mM.

5. 1,1′-Dioctadecyl-3,3,3′,3′-tetramethylindodicarbocyanine, 4-chlorobenzenesulfonate salt (DiD). Stocks are reconstituted in chloroform.

6. Lipid mix: 0.84 mM DOPC, 0.15 mM DOPS, 0.05 mM DOPIP2, and 0.01 mM DiD prepared in chloroform in a glass vial and stored at −80 °C.

7. Glass vials with PTFE liner.

8. Methanol.

9. Chloroform.

10. Phosphate buffered-saline (PBS) pH 7.4.

11. Fatty acid-free BSA.

12. Oxygen scavenger cocktail:

    (a) 100× glucose oxidase: 20 mg/ml in HBS, frozen as aliquots and stored at −80 °C.

    (b) 100× catalase: 3.5 mg/ml in HBS, frozen as aliquots and stored at −80 °C.

    (c) 100× glucose: 450 mg/ml (2.25 M) in water, filtered and stored at −20 °C.

    (d) 1000× DTT: 1 M DTT in water, added freshly before assay.

13. Assay buffer: Degassed HBS with 1× oxygen scavenger cocktail, added freshly before assay.

14. Indium tin oxide (ITO)-coated glass slides to fit FCS2 chamber.

2.5 Fluorescence Microscopy and Image Analysis

Inverted fluorescence microscope equipped with a 100× oil-immersion lens and a stable fluorescence LED light source. Fluorescence emission is collected through single-band pass filters with excitation/emission wavelength band passes of 562 ± 40 nm/624 ± 40 nm for Texas Red, and 628 ± 40 nm/692 ± 40 nm for DiD probes. Image acquisition is controlled by MetaMorph software (Molecular Devices). Image analysis is carried out on a workstation using Fiji [13] and Graphpad Prism (Graphpad Software Inc.).

3 Methodology

3.1 Expression and Purification of Epsin1

Carry out all procedures at room temperature, unless stated otherwise.

1. Rat Epsin1 (Uniprot ID: O88339; Isoform 1) is cloned into a pET15B vector with a 6xHis tag at the N-terminus and StrepII tag at the C-terminus and confirmed by sequencing (*see* **Note 1**).

2. To generate a starter culture, a freshly transformed colony of BL21 (DE3) cells is inoculated in 5 ml of LB broth with the recommended concentration of ampicillin and grown overnight at 37 °C.

3. To induce protein expression, 500 ml of autoinduction medium is inoculated with this starter culture and grown under shaking conditions for 30 h at 18 °C.

4. Cultures are spun at $4000 \times g$ for 10 min and the pellet is washed once with PBS and stored at −40 °C till further use.

5. A frozen bacterial pellet from a 500 ml culture is resuspended in 20 ml of HBS supplemented with 1 mM PMSF and lysed by sonication.

6. The lysate is spun at $18{,}000 \times g$ for 20 min at 4 °C and the supernatant is collected into a 50 ml Falcon.

7. 2 ml of packed volume of HisPur cobalt resin is poured into a PD-10 column and washed with 20 ml of HBS by connecting the column to a P-1 peristaltic pump.

8. The equilibrated resin is then poured into the Falcon containing the culture supernatant and incubated for 1 h at 4 °C under rocking conditions.

9. The supernatant is then poured back into the PD-10 column and washed with 50 column volumes of HBS, while triturating with a 1 ml micropipette.

10. Bound protein is eluted with 10 ml of HisPur cobalt resin elution buffer.

11. The elution is applied onto a StrepTrap HP column using the peristaltic pump.

12. The column is washed with 5 column volumes of HBS.

13. Bound protein is eluted in 10× 1 ml fractions of StrepTRAP elution buffer.

14. Elutions are run on a 10% SDS-PAGE.

15. Fractions containing pure Epsin1 are flash-frozen in liquid nitrogen and stored at −80 °C.

16. Epsin1 is dialyzed overnight against HBS and spun at 100,000 × g for 20 min to remove aggregates before use in microscopy experiments (*see* **Note 2**).

17. Protein concentration is estimated using UV absorption spectroscopy based on a molar extinction coefficient (ε) estimated using the ExPASy ProtParam tool.

3.2 Purification of Native Clathrin

Clathrin is extracted from clathrin-coated vesicles (CCVs) isolated from brain tissue as described earlier [14], with a few modifications.

1. Brain tissue (~110 g) (*see* **Note 3**) is first cleaned in a minimum volume of cold PBS to remove meninges and blood vessels.

2. The tissue is then chopped into small pieces and homogenized in an equal volume of cage assembly buffer using a Waring blender.

3. The homogenate is spun at 17,700 × g for 30 min at 4 °C.

4. The supernatant is collected and spun at 70,000 × g for 1 h at 4 °C.

5. The resultant pellet is resuspended in a minimum volume of cage assembly buffer using a Dounce homogenizer and mixed with an equal volume of floatation buffer.

6. The suspension is spun at 41,400 × g for 40 min at 4 °C.

7. The supernatant is collected and diluted 5-fold with cage assembly buffer supplemented with 0.1 mM PMSF and spun at 85,195 × g for 1 h at 4 °C.

8. The pellet (enriched in CCVs) is resuspended in a minimum volume of cage assembly buffer and added dropwise directly into liquid nitrogen. Frozen drops are then collected through a sieve, transferred to Falcon tubes and stored at −80 °C. Due to the large volumes involved, this approach of directly freezing samples saves on time and is convenient for downstream steps.

9. About 5–10 mg of CCV pellets is thawed in 5 ml of cage disassembly buffer and homogenized using a Dounce homogenizer.

10. Samples are left at room temperature for 2–3 h and spun at 100,000 × g for 1 h at room temperature.

11. The supernatant is collected, dialyzed overnight against cage disassembly buffer and spun at 100,000 × g to remove aggregates.

12. The precleared supernatant is loaded onto a Q Sepharose column, washed with 10 column volumes of cage disassembly buffer and eluted against a linear gradient of anion exchange elution buffer on an AKTA Prime Plus chromatograph setup.

13. 0.5 ml elutions are collected and run on a 10% SDS-PAGE.

14. Fractions containing a prominent band at ~180 kDa, representing clathrin, are pooled and mixed with saturated ammonium sulfate solution (final concentration = 30% (v/v)) and incubated for 1 h at 4 °C.

15. Samples are spun at 20,000 × g for 1 h at 4 °C and the pellet is resuspended in a minimum volume of cage disassembly buffer supplemented with 10% v/v glycerol, flash-frozen in liquid nitrogen and stored at −80 °C.

3.3 Fluorescent Labeling of Clathrin

1. A stock of the thiol-reactive Texas Red C2 maleimide dye is prepared in DMSO and added at tenfold molar excess to purified clathrin in cage disassembly buffer. Typical concentrations of the clathrin preparation used for labeling is ~0.8–1.0 μM.

2. The reaction is incubated for 1 h at room temperature and subsequently quenched by adding DTT to a final concentration of 1 mM.

3. The reaction is then dialyzed against cage disassembly buffer to remove free dye, which is confirmed by running the labeled clathrin preparation on a 10% SDS-PAGE gel, where free dye if any migrates with the dye front (see **Note 5**).

4. In order to enrich for assembly-competent labeled triskelia, labeled clathrin is dialyzed against cage assembly buffer overnight at 4 °C, spun at 100,000 × g to pellet cages and resuspended in cage disassembly buffer.

5. Labeled clathrin preparations are flash-frozen in liquid nitrogen and stored at −80 °C.

6. Clathrin concentration is estimated using UV absorption spectroscopy based on a molar extinction coefficient (ε) estimated using the ExPASy ProtParam tool (see **Note 4**).

3.4 Passivation of Glass Coverslips for Fluorescence Microscopy

In order to make SMrT templates (see below) and prevent nonspecific adsorption of proteins, glass coverslips are passivated by covalent attachment of polyethylene glycol (PEG) according to earlier reports [15] (see **Note 6**).

1. 40 mm glass coverslips are cleaned with 3 N NaOH for 5 min and rinsed extensively with deionized water.

2. Clean coverslips are treated with piranha solution for 1 h at room temperature and rinsed extensively with deionized water and dried on a heating block set to 90 °C.

3. Dry coverslips are arranged on a clean and dry glass petri dish and just enough neat 3-glycidyloxypropyltrimethoxysilane is poured over them to immerse every coverslip in the solution.

4. Coverslips are kept under vacuum for 5 h.

5. Silanized coverslips are dipped in acetone to remove excess silane, air-dried and immediately immersed into molten PEG8000 maintained at 90 °C.

6. Coverslips are left undisturbed for 48–60 h.

7. PEGylated coverslips are removed from the PEG solution, rinsed extensively with water and stored dry in a closed container.

3.5 Preparation of SMrT Templates

SMrT templates comprise an array of membrane tubes laid out on a passive surface. These templates are prepared according to published protocols [9, 10].

1. Lipid mix is brought out of storage at −80 °C (*see* **Note 7**). The mix is brought to room temperature before use.

2. A PEGylated coverslip is cleaned with 1% SDS, water, methanol, and then water and wiped dry with a soft tissue paper.

3. A small aliquot (~3 μl) of the 1 mM (total lipid) stock is spread as a thin band at 1/3 distance from the edge of the rectangular working area of the coverslip (*see* **Note 8**) (Fig. 1).

4. The coverslip is kept under high vacuum for 5 min to remove traces of chloroform.

5. A 35 μl flow cell is assembled by placing a 0.1 mm silicone spacer between the PEGylated coverslip and an ITO-coated glass slide in an FCS2 closed chamber that is attached to a low-binding eppendorf at the inlet (Fig. 1).

6. The eppendorf is filled with 1 ml of filtered and degassed PBS and with a micropipette suction is applied at the outlet to introduce buffer into the flow cell.

7. The buffer in left in the chamber undisturbed for 10 min at room temperature to allow complete hydration of the dried lipid.

8. The flow cell is connected to a microperfusion pump and flow rate is set to reach ~30 mm/s of particle velocity inside the chamber.

9. Enough buffer is flowed till all the excess lipid reservoir has been extruded into membrane tubes. SMrT templates are judged ready for experiments when the entire membrane reservoir is extruded into tubes that remain pinned at discrete sites to the surface (*see* **Note 9**).

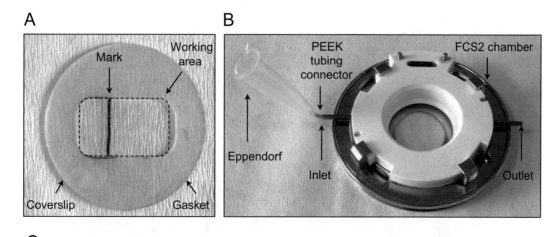

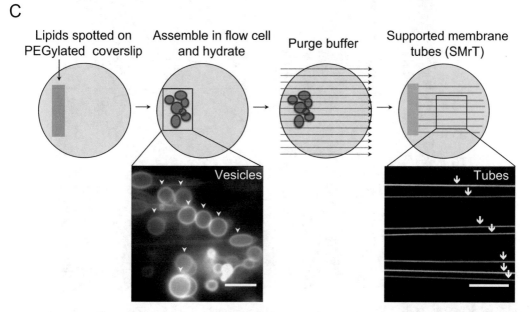

Fig. 1 (**a**) Photograph of a 40 mm glass coverslip placed on a silicone gasket showing the mark where the lipid mix is spotted. The rectangle with dotted line marks the working area of the chamber. (**b**) Photograph of the FCS2 chamber from Bioptechs connected at the inlet to a low-binding eppendorf through a PEEK tubing, which serves as the reservoir. The outlet is connected to a microperfusion pump. (**c**) Schematic of generation of SMrT templates. Hydration of lipids leads to the formation of vesicles (white arrowheads) that are subsequently extruded by buffer flow into long membrane tubes (white arrows) that span the entire working area of the flow cell. Scale bars = 10 μm

3.6 Clathrin Assembly on Epsin1-Coated SMrT Templates

1. Immediately following 3.6, 1 ml of HBS containing 1% (w/v) fatty acid-free BSA is flowed onto the SMrT template preparation and incubated for 10 min at room temperature (*see* **Note 10**).

2. Excess BSA is washed off by passing 200 μl of assay buffer.

3. 200 μl of 0.2 μM purified Epsin1 suspended in assay buffer is then flowed in and incubated for 10 min.

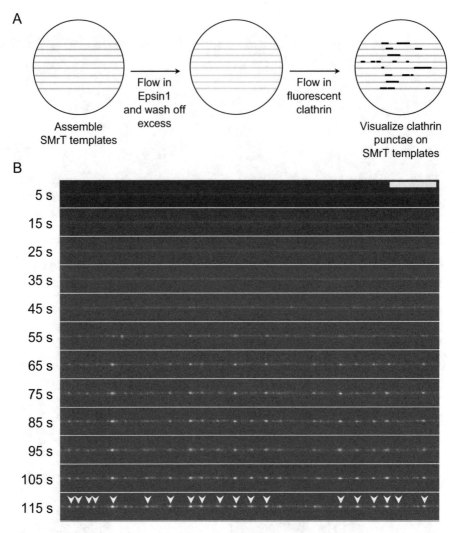

Fig. 2 (**a**) Schematic of the SMrT assay to monitor clathrin assembly. (**b**) A montage showing clathrin assemblies (white arrowheads) formed with time on a single membrane tube. Scale bar = 10 μm

4. Excess Epsin1 is washed off by passing 600 μl of assay buffer.

5. The eppendorf is filled with 200 μl of a 40 nM solution of Texas Red-labeled clathrin that is freshly diluted in assay buffer (*see* **Note 11**).

6. An appropriate field of view is selected and imaging parameters for acquisition are configured (*see* **Note 12**).

7. The assay is initiated by turning on the perfusion pump and simultaneously acquiring a time-lapse sequence (*see* **Note 13** and Fig. 2).

8. After acquisition of the required frames, the time-lapse sequence is saved for further analysis.

9. The pump is stopped and the flow cell is rinsed sequentially with 1 ml each of deionized water, 1% SDS, deionized water, 70% ethanol, and deionized water.

3.7 Image Analysis

Analysis of kinetics of clathrin assembly constitutes the following steps.

1. The dead time of the flow cell is estimated in situ for each experiment by calculating the onset of fluorescent clathrin into the microscope field using a plateau followed by one-phase exponential rise function. Frames representing the dead time of the flow cell are removed from the original time-lapse sequences.

2. Time-lapse sequences are then condensed to reduce them to easily manageable files. We typically reduce time-lapse sequences by a factor of 50 which increases the time interval between frames to 5 s.

3. Using the "*Multiple Kymograph*" tool in Fiji, a "*Test*" kymograph is generated of clathrin polymerization on SMrT templates by placing a line across the membrane tube (*see* Fig. 3). Another "*Background*" kymograph is generated by moving the line to the background area of the time-lapse sequence.

4. Using the "*Image Calculator*" tool in Fiji, the "*Background*" kymograph is subtracted from the "*Test*" kymograph to generate a "*Background-corrected*" kymograph.

5. Using the "*Image to Results*" tool in Fiji, pixel intensity versus time data is extracted for all pixels on the "*Background-corrected*" kymograph.

6. Pixel intensity versus time data is pasted into Graphpad Prism and fitted to a plateau followed by one-phase exponential rise function. Fits with an $R^2 \geq 0.8$ (*see* **Note 14**) are selected to provide an estimate of the onset, rate, and extent of clathrin assembly.

4 Notes

1. A tandem affinity purification protocol ensures purification of full length protein avoiding contamination from truncated forms. This is particularly important for Epsin1 since the unstructured C-terminal tail, which contains clathrin binding sites, is prone to proteolysis.

2. Epsin1 aggregates bound to the membrane can change the apparent kinetics and extent of clathrin polymerization.

3. We extract clathrin from goat brain tissue. Fresh brains provide better yields of clathrin and we recommend a visit to the local abattoir before the experiment.

172   Devika Andhare et al.

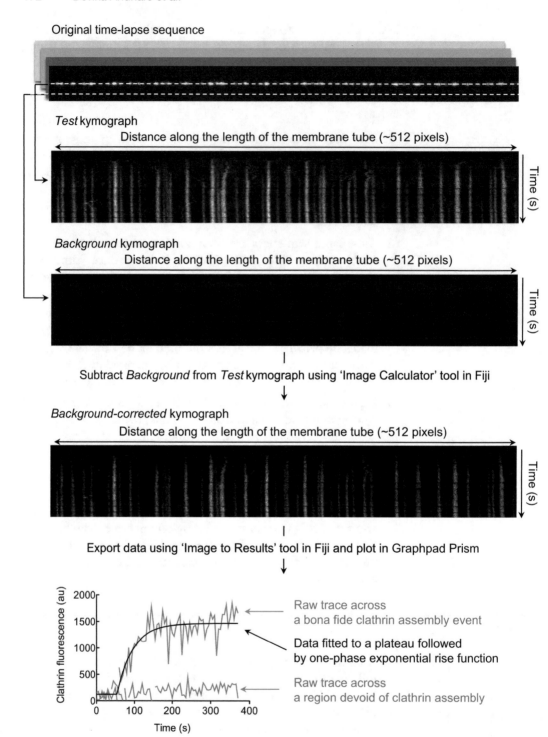

Fig. 3 Image analysis methodology to extract kinetics of clathrin assembly from kymographs

4. Clathrin concentration is calculated assuming a 1:1 stoichiometry of clathrin heavy and light chains. The molar extinction coefficient (ε) is calculated to be 236,240 M/cm at 280 nm using the amino acid sequences of the human clathrin heavy chain (CHC17) and human clathrin light chain, isoform A (CLTA) deposited in Uniprot.

5. Depending on its hydrophobicity, free/unreacted dye present in the labeled protein preparation can interact with membranes [16]. It is therefore important to ensure that the labeled protein preparation is removed of any free dye. This can be tested by running it on a SDS-PAGE gel before use in experiments since any free dye migrates along with the dye front.

6. Tips on PEGylation and repeated use of PEGylated glass coverslips:

   (a) To ensure removal of residual NaOH, rinse with minimum volume of concentrated H_2SO_4 before piranha treatment.

   (b) A good indicator of effective piranha treatment is that upon washing coverslips a film of water is retained on the glass surface.

   (c) Do not extend the piranha treatment beyond an hour. Prolonged piranha treatment makes the coverslips brittle and difficult to handle.

   (d) GOPTS can react with water causing inefficient silanization of coverslips. It is therefore important to dry coverslips and the container used for silanization beforehand.

   (e) PEG8000 takes a while to melt at 90 °C so prepare beforehand.

   (f) PEG8000 dissolves easily in hot water. After PEGyation rinse extensively with hot water to remove excess PEG8000.

   (g) Long chain PEG has been reported to interact with SDS [17]. This interaction could retain residual SDS on the PEGylated glass surface during washes, which can be removed by a subsequent wash with methanol.

   (h) Coverslips can be reused four to five times without significant loss of surface passivation if cleaned sequentially with 1% SDS, water, methanol, and water before and in between experiments.

7. Epsin1 displays high affinity for PIP_2 [9] and serves to recruit it to SMrT templates. DiD is a fluorescent lipid analog to enable visualization of membranes under the fluorescence microscope.

8. Spotting the lipid too close to the inlet leads to improper extrusion of the membrane reservoir into membrane tubes.

9. The membrane tubes are initially suspended in buffer and with time settle down and get pinned to the coverslip surface.

10. Despite the glass surface being passivated by covalent attachment of PEG8000, clathrin displays a strong tendency to adsorb to the surface, which is significantly reduced by coating the coverslips surface with fatty acid-free BSA.

11. Clathrin self-assembles upon exposure to buffer of acidic pH (around 6.8). It is therefore important to maintain the pH of the assay buffer.

12. We typically record for 4000 frames at 0.1 s time interval per frame.

13. Arrival of clathrin into the field is apparent by an increase in background fluorescence. Shortly afterward, clathrin punctae start appearing on the epsin-coated membrane tube.

14. The $R^2 \geq 0.8$ is a stringent criterion that will sort out artifacts caused by microscope focus drifts and/or flow-induced lateral movement of fluorescent clathrin punctae.

Acknowledgment

This work was supported by research grants from the Wellcome Trust-Department of Biotechnology (DBT) India Alliance and intramural funding from IISER Pune. T.J.P. was a Senior Fellow of the Wellcome Trust-Department of Biotechnology (DBT) India Alliance. D.A thanks the University Grants Commission, and S.S.H thanks the Council for Scientific and Industrial Research (CSIR) for graduate research fellowships. We thank members of the Pucadyil lab for comments on the manuscript.

References

1. Traub LM (2009) Tickets to ride: selecting cargo for clathrin-regulated internalization. Nat Rev Mol Cell Biol 10:583–596

2. Mcmahon HT, Boucrot E (2011) Molecular mechanism and physiological functions of clathrin-mediated endocytosis. Nat Rev Mol Cell Biol 12:517–533

3. Traub LM (2005) Common principles in clathrin-mediated sorting at the Golgi and the plasma membrane. Biochim Biophys Acta 1744:415–437

4. Kozik P et al (2012) A human genome-wide screen for regulators of clathrin-coated vesicle formation reveals an unexpected role for the V-ATPase. Nat Cell Biol 15:50–60

5. Dannhauser PN, Ungewickell EJ (2012) Reconstitution of clathrin--coated bud and vesicle formation with minimal components. Nat Cell Biol 14:634–639

6. Kelly B et al (2014) AP2 controls clathrin polymerization with a membrane-activated switch. Science 34:458–463

7. Ford MGJ et al (2002) Curvature of clathrin-coated pits driven by epsin. Nature 419:361–366

8. Saleem M et al (2015) A balance between membrane elasticity and polymerization energy sets the shape of spherical clathrin coats. Nat Commun 6:1–10

9. Holkar S et al (2015) Spatial control of epsin-induced clathrin assembly by membrane curvature. J Biol Chem 290:14267–14276

10. Dar S et al (2015) A high-throughput platform for real-time analysis of membrane fission

reactions reveals dynamin function. Nat Cell Biol 17:1588–1596

11. Dar S et al (2017) Use of the supported membrane tube assay system for real-time analysis of membrane fission reactions. Nat Protoc 12:390–400

12. Pucadyil TJ, Holkar SS (2016) Comparative analysis of adaptor-mediated clathrin assembly reveals general principles for adaptor clustering. Mol Biol Cell 27:3156–3163

13. Schindelin J et al (2012) Fiji: an open source platform for biological image analysis. Nat Methods 9:676–682

14. Campbell C et al (1984) Identification of a protein kinase as an intrinsic component of rat liver coated vesicles. Biochemistry 23:4420–4426

15. Piehler J et al (2000) A high-density poly(ethylene glycol) polymer brush for immobilization on glass-type surfaces. Biosens Bioelectron 15:473–481

16. Hughes L et al (2014) Choose your label wisely: water-soluble fluorophores often interact with lipid bilayers. PLoS One 9(2):e87649

17. Dai S, Tam KC (2001) Isothermal titration calorimetry studies of binding interactions between polyethylene glycol and ionic surfactants. J Phys Chem B 105:10759–10763

Chapter 13

Real-Time Monitoring of Clathrin Assembly Kinetics in a Reconstituted System

Jeffery Yong, Yan Chen, and Min Wu

Abstract

Clathrin-coated pits (ccp) are important structures that cells use for internalizing materials and regulating plasma membrane homeostasis. We had previously described an assay of reconstituting ccp assembly on sheets of basal plasma membranes. Here, we describe a workflow to adapt this system for monitoring the assembly of ccps over time using total internal reflection fluorescence (TIRF) microscopy.

Key words Endocytosis, Reconstitution, Clathrin, Cell-free, TIRF, Fluorescent imaging, Budding, Single vesicle

1 Introduction

Live imaging of endocytic process at single vesicle resolution has transformed our understanding of this dynamic machinery [1–3]. In recent years, exciting advances have been made in the area of real-time imaging of clathrin dynamics with the introduction of combined TIRF and wide-field illumination [4], superresolution imaging by stochastic optical reconstruction microscopy [5], structured illumination microscopy [6, 7], and genome-edited cell lines [8]. Accumulating evidence suggests endocytic processes are dynamic and heterogeneous with variable lifetime of clathrin assembly which could be regulated by cargo [9], lipid metabolism [10], and accessory proteins [11, 12]. However, mechanistic understanding of the kinetic regulation of clathrin assembly in vivo remains challenging.

Delineating complex processes require in vitro assays where experimental conditions could be precisely controlled. The assembly of endocytic pits on the membrane template has been successfully reconstituted in vitro using several model systems. Many of these systems allow high-resolution ultrastructure studies by elec-

The original version of this chapter was revised. A correction to this chapter can be found at
https://doi.org/10.1007/978-1-4939-8719-1_19

Laura E. Swan (ed.), *Clathrin-Mediated Endocytosis: Methods and Protocols*, Methods in Molecular Biology, vol. 1847,
https://doi.org/10.1007/978-1-4939-8719-1_13, © Springer Science+Business Media, LLC, part of Springer Nature 2018

tron microscopy [13, 14] but not real-time monitoring of clathrin assembly at single-vesicle resolution. We have successfully reconstituted clathrin-coated pit formation on the plasma membrane sheets isolated from adherent cells [15]. The planar geometry of the substrate is ideally suitable for high-resolution, real-time imaging of the assembly process by methods such as TIRF microscopy. At the same time, the open, cell-free system allows for easy manipulation of cytoplasmic protein composition. Previously we have described the detailed protocol for the cell-free reconstitution assay [16]. In the following chapter we focus on how to set up the assay for real-time kinetics studies using a perfusion system that allows for the controlled initiation of ccp assembly and the high-resolution TIRF microscopy observation of the assembly process.

2   Materials

2.1   Buffers

1. Cytosolic buffer: 25 mM 4-(2-hydroxyethyl)-1-piperazineethanesulfonic acid (HEPES) pH 7.4, 120 mM potassium glutamate, 20 mM potassium chloride, 2.5 mM magnesium acetate, 5 mM ethylene glycol-bis(β-aminoethyl ether)-N,N,N',N'-tetraacetic acid (EGTA), 1 mM dithiothreitol (DTT, added fresh). Store at 4 °C.

2. Coating buffer: 0.1 M sodium borate pH 8.5.

3. Phosphate buffered saline (PBS).

4. Stripping buffer: 0.5 M Tris–HCl pH 7.0 (measured at 37 °C).

2.2   Reagents (Stock Concentrations Stated)

1. 1M hydrochloric acid (HCl).

2. 2 mg/mL poly-D-lysine (PDL; 100×), dissolved in coating buffer. Store at −20 °C.

3. 100 mM ATP (100×), dissolved in deionized water. Store at −20 °C.

4. 668 U/mL creatine phosphokinase: (40×), dissolved in PBS. Store at −80 °C as single-use 20 μL aliquots to avoid repeated freeze-thawing. (*see* **Note 1.**)

5. 668 mM phosphocreatine: (40×), dissolved in deionized water. Store at −20 °C.

6. 6 mM GTPγS (40×), dissolved in deionized water. Store at −20 °C.

2.3   Cell Cultures and Media

1. PTK2 cells (ATCC).

2. 20 mm diameter coverslips.

3. 35 mm sterile tissue culture dishes.

4. Crystallizing dish.

2.4 Purified Cytosolic Extracts and Fluorescent Clathrin

1. Cytosolic extract is prepared from fresh whole brain as previously described [16]. Porcine and rodent (Mus and Rattus) brain cytosolic extracts have been used to assemble ccp successfully. The aliquots of cytosol (150 μL each) should be flash-frozen in liquid nitrogen and stored at −80 °C. Avoid freeze–thaw cycles that may cause degradation of enzymatic activity.

2. Clathrin is purified based on previously reported protocols from fresh porcine or bovine brains [17]. Purified clathrin triskelia are labeled using the maleimide derivative of Alexa dyes (*see* **Note 2.**).

2.5 Equipment

1. Probe sonicator
   A sonicator with a small probe diameter is used to disrupt the PTK2 cells and leave the basal plasma membranes intact on the glass coverslips. We use a Vibra-Cell VC-505 Ultrasonic processor (Sonics and Materials Inc.) with a 1/8″ tapered microtip probe.

2. Perfusion system
   The perfusion system allows for the controlled initiation of ccp assembly and the observation of the entire assembly process. We use a customized Chamlide CF (Live Cell Instrument) perfusion chamber, which has a fluid channel volume of approximately 4 μL and a total dead volume of around 50 μL. These small volumes are advantageous for minimizing the amounts of reagents that are required for each experiment. The perfusion chamber has two separate inlets and one outlet (Fig. 1). The inlets are connected to buffer reservoirs with motorized valves that are controlled by MetaMorph imaging software. The outlet is attached to a PHD Ultra syringe pump (Harvard Apparatus) that is triggered by MetaMorph imaging software to withdraw fluid out of the perfusion chamber. Finally, tubing that function as sample loops are situated between the buffer reservoirs and the imaging chamber inlets. We typically introduce samples into the chamber at a flow rate of 3.3 μL per second for a total volume of 100 μL. High flow rates may damage the membrane sheets.

3. Imaging system
   We routinely use TIRF microscopy for the temporal observation of ccp assembly. As the membrane sheets are largely flat and close to the coverslip, they are easily within the range of the TIRF evanescent wave. Another advantage of using TIRF is that soluble fluorescent entities are out of range of the evanescent wave and therefore contribute very little background noise. The perfusion chamber and sample loops are fitted into a holder within a heated stage incubator on a Nikon ECLIPSE Ti-S inverted microscope with iLas2 TIRF illumination source.

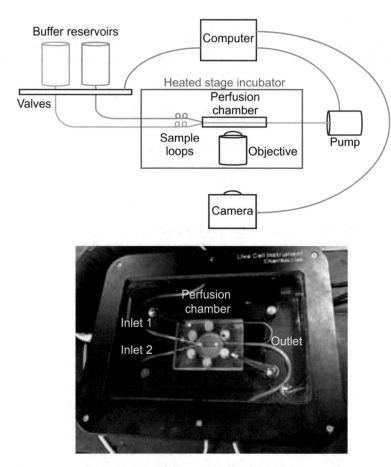

Fig. 1 Perfusion system for observing ccp assembly. Top panel, schematic diagram of the perfusion and imaging system used. Colored lines trace the pathway of fluid through the entire system. Bottom panel, photograph of the perfusion chamber and sample loops within the heated stage incubator of the microscope. The inlet and outlet tubing for the perfusion chamber is as indicated

The illumination source, microscope stage, and Evolve EMCCD camera are all under MetaMorph software control (Fig. 1).

3   Methods

3.1   Preparation of Coverslips

Step 1: Coverslip coating (prepared 2–4 days in advance of plating).

Coverslips are cleaned by soaking them with 1 M HCl for 15 min at room temperature, followed by three rinses in deionized water. The coverslips are then sterilized and stored in ethanol until required. These can be made 2–4 days in advance of the experiment. The day before the experiment, the coverslips are dried in sterile 35 mm tissue culture dishes within a sterile lamina flow hood. They are then rinsed with sodium borate buffer and air dried briefly. The coverslips are coated by adding 150 μL of poly-D-lysine (diluted in coating buffer) for 30 min in a 37 °C humid tissue culture incubator. After coating, the coverslips are rinsed twice and soaked in sterile deionized water for at least 24 h before cell plating.

Step 2:   Plating of PTK2 cells (2–4 days before sonication and imaging experiment)

PTK2 cells are used to generate the membrane sheets on which the ccp are to be assembled. They are maintained in minimum essential medium supplemented with 10% fetal bovine serum and are grown in a humid 5% CO_2 atmosphere at 37 °C. The water in the dishes containing PDL coated coverslips is removed and the coverslips are air dried briefly in a sterile lamina flow hood. Cells are seeded and incubated at 37 °C, 5% CO_2 incubator for 2–4 days. The final density of the cells is an important factor for the generation of good basal membrane sheets. For PTK2 cells, we typically seed 0.6×10^6 cells per 35 mm diameter dish. The cells are then grown for 2–4 days to allow them to form an even and confluent monolayer before being used for producing the basal membrane sheets. The final monolayer contains between 450 and 550 cells per mm².

3.2 Sample Preparation on the Day of Imaging

An overview of the assay is summarized in Fig. 2. The specific details are provided below.

Step 1:   Cleaning and preparation of the perfusion system and imaging chamber.

The buffer reservoirs and tubing of the entire perfusion system are rinsed and filled with freshly made cytosolic buffer. The system with buffer is then allowed to equilibrate to room temperature prior to the loading of the coverslips containing the membrane sheets and cytosolic extracts.

Step 2: Sonication and generation of membrane sheets.

Tissue culture medium is removed and the entire dish with coverslip is immersed into a crystallization dish containing cold cytosolic buffer with the probe positioned 20 mm above the dish and coverslip. The cells are sonicated using a 1 s pulse at 25% of maximum power of the sonicator. Further pulses and repositioning may be necessary to ensure complete disruption in areas not directly under the probe tip. The raw basal membranes are then rinsed in cytosolic buffer and are usually used within 30 min of sonication (*see* **Note 3**).

Step 3: Stripping of endogenous structures.

Mechanical disruption by sonication removes the apical parts of the cells and endogenous clathrin structures would remain on the raw membrane sheets. We have observed numerous clathrin coated pits in various stages of assembly on these raw membrane

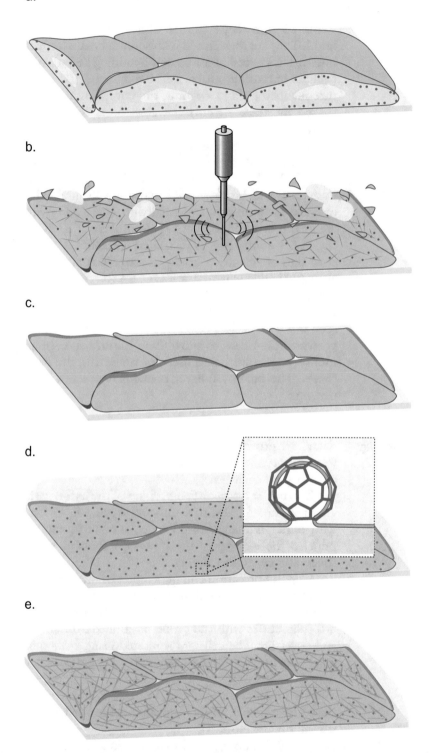

Fig. 2 Schematic overview of the assembly assay. (**a**) Cells are plated to form an even monolayer. (**b**) Cells are disrupted by sonication, leaving basal membranes and associated materials. (**c**) Endogenous/residual materials are stripped away by incubation with Tris–HCl. (**d**) Ccp assembles upon the introduction of the reaction mix. Insert depicts a single ccp with the clathrin cage in red. (**e**) Late stages of ccp assembly with rampant actin polymerization

sheets, similar to what has been reported [18, 19]. This endogenous clathrin could be disassembled by incubating the membrane with stripping buffer [20]. This stripping step is necessary for removing existing ccp and ensuring that subsequent incubation of cytosolic extracts would produce de novo ccp (*see* **Note 4**.). Cytosolic buffer from the dishes is removed and the dish filled with stripping buffer and incubated at 37 °C for 3 min. The stripping buffer is then removed and the dishes rinsed twice with fresh cytosolic buffer.

Step 4: Loading of coverslips in preparation for imaging.

The upper section of the perfusion chamber is inverted and filled with cytosolic buffer to form a droplet over the channel of the perfusion chamber. This step is important to prevent bubbles of air from being trapped within the perfusion chamber. The coverslip containing the membrane sheets is loaded into the coverslip holder of the perfusion chamber and assembled onto the inverted upper section of the perfusion chamber. Excess cytosolic buffer is then wiped off and the exposed bottom surface of the coverslip cleaned using 70% ethanol.

Step 5: Preparation and loading of reaction mix into the sample loops.

The typical reaction mix per imaging experiment is composed of 100 μL of 4 mg/mL of cytosolic extract mixed with 0.02–0.04 nM of fluorescently labeled clathrin per μg of cytosolic extract, 0.15 mM GTPγS, and an ATP regenerating system (consisting of 1 mM ATP, 16.7 U/mL creatine phosphokinase, and 16.7 mM phosphocreatine).

Each reaction mixture is prepared just prior to use and immediately loaded into the sample loops between the perfusion chamber and the buffer reservoirs. A simple way of loading the reaction mix is by detaching the tubing where it enters the perfusion chamber and injecting the mix into the tubing via a 100 μL pipette with the corresponding reservoir valve open. The valve is then closed, the pipette removed and the tubing reattached to the perfusion chamber. Care should be taken to avoid introducing any bubbles during the entire process. It must be noted that reaction mix loading should be done after loading the coverslips to minimize accidental introduction of the reaction mix onto the membrane sheets during handling. The entire perfusion chamber and sample loop assembly are then incubated for approximately 5 min in the stage-top incubator to equilibrate their temperature to 37 °C and also to minimize thermal drifting during imaging.

Step 6:   Time-lapse TIRF microscopy.

A drop of optical immersion oil is applied to the objective and appropriate adjustments to the correction ring of the objective (if present) are made. The membrane sheets are then brought into focus, either by fluorescent membrane markers or by differential interference contrast microscopy. The excitation laser angle is then calibrated for TIRF illumination (*see* **Note 5**.). Ccp assembly is initiated by opening the appropriate valve and the perfusion of the reaction mix from the sample loops into the imaging chamber. The typical sequence of TIRF imaging is as follows:

1. Preinitiation imaging (10 frames at 0.2 Hz) is acquired for determining the background level before ccp assembly.

2. Opening of valve and initiation of perfusion by the syringe pump.

3. Closing valve after sample perfusion is complete, usually 30 s after initiation of the pump.

4. Imaging until target duration is achieved.

The typical time-lapse sequence of ccp assembly is shown in Fig. 3a. The exposure time is determined based on optimal signal-to-noise ratio, usually approximately 100 ms per exposure. The frame rate for a typical assembly assay is 1–5 s per frame (0.2–1.0 Hz) for a total observation time of 5–15 min (*see* **Notes 6** and **7**.).

4 Notes

1. We have observed reduced activity after extended storage of creatine phosphokinase solutions and thus recommend that the enzyme should be used within 6 months of preparation.

2. We label purified clathrin triskelia using maleimide-Alexa dyes. Only preparations of conjugated clathrin that can undergo a round of assembly and disassembly to verify its functionality are stored and used for experiments [15]. Maleimide derivative of Alexa dyes including Alexa 488, Alexa 555, and Alexa 647 have been used to conjugate clathrin successfully.

3. Care should be taken to prevent the drying out of the membranes on the coverslips. Accidental drying of areas on the coverslip would cause the membranes there to be irreversibly damaged and no longer appear as a sheet.

4. Using membrane derived from cells expressing clathrin-light chain-GFP and cytosol doped with Alexa 647 clathrin, we confirmed incorporation of Alexa 647 clathrin in the cytosol in both preexisting and de novo assembled clathrin-coated structures. Removing the preexisting clathrin structures are not essential for reconstitution of membrane budding or fission reaction but could reduce heterogeneity of the early assembly reaction. We treated the membrane sheets with Tris–HCl to strip off preexisting clathrin on the membranes because the amine groups on Tris are expected to break the high-affinity salt bridges between two pairs of lysine–aspartic acid residues from two antiparallel

clathrin heavy chains [21]. This procedure allows us to study exclusively the de novo ccp formation induced by the cytosol.

5. Effect of TIRF penetration depth. We could observe and quantify ccp assembly in a wide range of TIRF field depths. Under most of the TIRF depths, two phases could be observed. The first phase is the assembly phase wherein the intensity of labeled clathrin increases until a peak or stable plateau is reached. This phase may represent the actual accumulation and assembly of the clathrin coat around the budding membrane. The second phase involves an apparent disappearance of the ccp shortly after its intensity peaks. Due to the inclusion GTPγS in our reaction mix, the budding vesicles do not pinch off but invaginate upward, possibly due to actin polymerization around the bud neck (Fig. 2). We confirmed that the decreasing phase is not due to the disassembly of the clathrin coat but due to the exiting of the clathrin out of the evanescent wave field by using sequential imaging of the assembly at two different TIRF excitation angles (Fig. 3a, b). Clathrin coats are still visible at higher penetration depths (300 nm in Fig. 3a, b) or in the wide field epifluorescence although it disappeared from the shallow penetration depth (100 nm in Fig. 3a, b). The kinetics of the decay phase, but not the assembly phase, is sensitive to the penetration depth of TIRF field (Fig. 3b, c). Thus, depending on the primary goal of the experiment, the penetration depth should be carefully considered and optimized.

6. Evaluating phototoxicity.
Fluorescent imaging produces oxidative stress and is inherently damaging to fluorophores, resulting in a gradual but significant loss of signal. In addition, oxidative stress also changes the properties of the membrane and could potentially inhibit the reconstitution reaction. It is necessary to assess the effects of various imaging parameters (laser power, acquisition time, interval, etc.) before standardizing your experimental condition. We find it convenient to use the peak time of clathrin assembly as a quick reference to determine whether ccp assembly is affected in our assay. In general, short laser wavelengths and high laser power generate oxidative stresses, which could inhibit the plateau of ccp assembly (deep invagination of the membrane), and to a lesser extent, the initiation of ccp assembly. Thus, we prefer to use Alexa647-conjugated clathrin illuminated using a 642 nm wavelength laser in a majority of our ccp assembly assays because it is the most robust probe in terms of its assembly kinetics, compatible with a wide range of laser power and exposure settings.

7. Oxygen scavenger system.
The inclusion of an oxygen scavenger system, such as the Glucose oxidase-Catalase system, minimizes phototoxicity

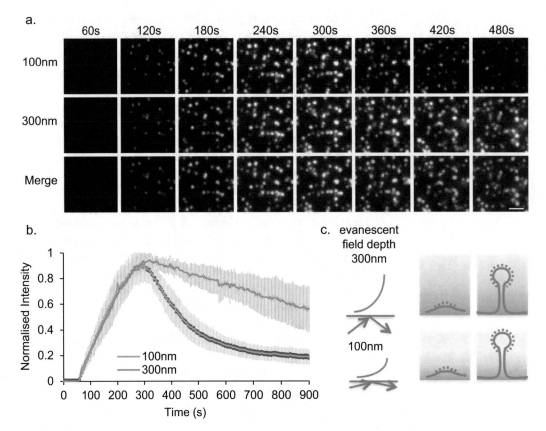

Fig. 3 Temporal observation of clathrin assembly at two different TIRF penetration depths. (**a**) Selected images from a typical ccp assembly imaged using two different evanescent wave penetration depths of 100 nm and 300 nm. Scale bar: 2 μm. (**b**) Plot of fluorescence intensity of the region of interests as observed at 100 nm and 300 nm TIRF penetration depths with time. Error bars: standard deviation. $n = 18$. (**c**) Schematic depicting disappearance of pits from shallow TIRF depth but the same pits remain in the deep evanescent wave depth

effects and should be included when a fluorescent lipid marker is used for imaging, or when non-laser source of excitation is used. However, this oxygen scavenger system alters the pH of the solution and should be used with caution [22]. Under our standard experimental condition where Alexa 647 clathrin assembly is monitored in the closed perfusion chamber, we noticed that inclusion of oxygen scavenger system is not always necessary.

Acknowledgment

This work was supported by the National Research Foundation (NRF) Singapore under its NRF Fellowship Program (M.W., NRF Award No: NRF-NRFF2011-09).

References

1. Gaidarov I, Santini F, Warren RA, Keen JH (1999) Spatial control of coated-pit dynamics in living cells. Nat Cell Biol 1:1–7. https://doi.org/10.1038/8971

2. Merrifield CJ, Feldman ME, Wan L, Almers W (2002) Imaging actin and dynamin recruitment during invagination of single clathrin-coated pits. Nat Cell Biol 4:691–698. https://doi.org/10.1038/ncb837

3. Taylor MJ, Perrais D, Merrifield CJ (2011) A high precision survey of the molecular dynamics of mammalian clathrin-mediated endocytosis. PLoS Biol 9:e1000604. https://doi.org/10.1371/journal.pbio.1000604

4. Saffarian S, Kirchhausen T (2008) Differential evanescence nanometry: live-cell fluorescence measurements with 10-nm axial resolution on the plasma membrane. Biophys J 94:2333–2342. https://doi.org/10.1529/biophysj.107.117234

5. Jones SA, Shim S-H, He J, Zhuang X (2011) Fast, three-dimensional super-resolution imaging of live cells. Nat Methods 8:499. https://doi.org/10.1038/nmeth.1605

6. Fiolka R, Shao L, Rego EH et al (2012) Time-lapse two-color 3D imaging of live cells with doubled resolution using structured illumination. Proc Natl Acad Sci 109:5311. https://doi.org/10.1073/pnas.1119262109

7. Li D, Shao L, Chen BC et al (2015) Extended-resolution structured illumination imaging of endocytic and cytoskeletal dynamics. Science 349:aab3500–aab3500. https://doi.org/10.1126/science.aab3500

8. Doyon JB, Zeitler B, Cheng J et al (2011) Rapid and efficient clathrin-mediated endocytosis revealed in genome-edited mammalian cells. Nat Cell Biol 13:331–337. https://doi.org/10.1038/ncb2175

9. Puthenveedu MA, Zastrow von M (2006) Cargo regulates clathrin-coated pit dynamics. Cell 127:113–124. https://doi.org/10.1016/j.cell.2006.08.035

10. Nakatsu F, Perera RM, Lucast L et al (2010) The inositol 5-phosphatase SHIP2 regulates endocytic clathrin-coated pit dynamics. J Cell Biol 190:307–315. https://doi.org/10.1083/jcb.201005018

11. Loerke D, Mettlen M, Yarar D et al (2009) Cargo and dynamin regulate clathrin-coated pit maturation. PLoS Biol 7:e57. https://doi.org/10.1371/journal.pbio.1000057

12. Saffarian S, Cocucci E, Kirchhausen T (2009) Distinct dynamics of endocytic clathrin-coated pits and coated plaques. PLoS Biol 7:e1000191. https://doi.org/10.1371/journal.pbio.1000191

13. Dannhauser PN, Ungewickell EJ (2012) Reconstitution of clathrin-coated bud and vesicle formation with minimal components. Nat Cell Biol 14:634–639f. https://doi.org/10.1038/ncb2478

14. Saleem M, Morlot S, Hohendahl A et al (2015) A balance between membrane elasticity and polymerization energy sets the shape of spherical clathrin coats. Nat Commun 6:1–10. https://doi.org/10.1038/ncomms7249

15. Wu M, Huang B, Graham M et al (2010) Coupling between clathrin-dependent endocytic budding and F-BAR-dependent tubulation in a cell-free system. Nat Cell Biol 12:902–908. https://doi.org/10.1038/ncb2094

16. Wu M, De Camilli P (2012) Supported native plasma membranes as platforms for the reconstitution and visualization of endocytic membrane budding. Methods Cell Biol 108:3–18. https://doi.org/10.1016/B978-0-12-386487-1.00001-8

17. Keen JH, Willingham MC, Pastan IH (1979) Clathrin-coated vesicles - isolation, dissociation and factor-dependent reassociation of clathrin baskets. Cell 16:303–312

18. Collins A, Warrington A, Taylor KA, Svitkina T (2011) Structural organization of the actin cytoskeleton at sites of clathrin-mediated endocytosis. Curr Biol:1–9. https://doi.org/10.1016/j.cub.2011.05.048

19. Heuser J (1989) Effects of cytoplasmic acidification on clathrin lattice morphology. J Cell Biol 108:401–411

20. Mahaffey DT, Peeler JS, Brodsky FM, Anderson RG (1990) Clathrin-coated pits contain an integral membrane protein that binds the AP-2 subunit with high affinity. J Biol Chem 265:16514–16520

21. Ybe JA (1998) Clathrin self-assembly is regulated by three light-chain residues controlling the formation of critical salt bridges. EMBO J 17:1297–1303. https://doi.org/10.1093/emboj/17.5.1297

22. Shi X, Lim J, Ha T (2010) Acidification of the oxygen scavenging system in single-molecule fluorescence studies: in situ sensing with a ratiometric dual-emission probe. Anal Chem 82:6132–6138. https://doi.org/10.1021/ac1008749

Chapter 14

Stimulated Emission Depletion (STED) Imaging of Clathrin-Mediated Endocytosis in Living Cells

Francesca Bottanelli and Lena Schroeder

Abstract

The recent development of probes and labeling strategies for multicolor super-resolution imaging in living cells allows cell biologists to follow cellular processes with unprecedented details. Here we describe how to image endocytic events at the plasma membrane of living cells using commercial (Leica, Abberior Instruments) or custom built STED microscopes.

Key words STED, Super-resolution imaging, Clathrin, Transferrin receptor, Endocytosis, Halo tag, SNAP tag

1 Introduction

Since the introduction of STED microscopy [1, 2], cell biologists have patiently waited for multicolor superresolution imaging to become easily accessible for imaging intracellular structures with nanoscale resolution in living cells. Commercial STED microscopes have become more common within the last few years, increasing the need for robust labeling strategies that work for a variety of subcellular structures.

Select organic dyes that are compatible with STED microscopy have been the preferred choice for labeling fixed cells. However, most of these dyes are not cell permeable. While the green/yellow fluorescent protein (FP) pair has been used for live STED imaging, FPs have poor photostability and the highly overlapping emission spectra require postprocessing [3]. Until recently, this has limited most STED applications to a small subset of extracellular live cell imaging applications and to fixed and permeabilized cells. Excitingly, two cell-permeable and STED compatible dyes, silicon-rhodamine (SiR) [4] and ATTO590, have been identified as a suit-

The original version of this chapter was revised. A correction to this chapter can be found at https://doi.org/10.1007/978-1-4939-8719-1_19

Laura E. Swan (ed.), *Clathrin-Mediated Endocytosis: Methods and Protocols*, Methods in Molecular Biology, vol. 1847, https://doi.org/10.1007/978-1-4939-8719-1_14, © Springer Science+Business Media, LLC, part of Springer Nature 2018

able dye-pair for STED imaging in living cells and the universality of the labeling strategy has been demonstrated with a variety of examples [5]. By using SiR and ATTO590 dyes with the self-labeling Halo-tag and SNAP-tag systems, we can now image clathrin with a variety of cargoes and machinery components. This will allow investigators to determine the nanoscale organization of components of the clathrin-mediated endocytic machinery in living cells, thus avoiding fixation and permeabilization artifacts. Here we present a general protocol for live-cell two-color STED imaging of clathrin-coated vesicles and endocytosed transferrin receptor at the plasma membrane. This protocol is easily adapted for other live-cell imaging applications.

2 Materials

1. 35 mm glass bottom dishes No 1.5 (MatTek or equivalent).
2. 1 M KOH (potassium hydroxide) solution.
3. Purified human plasma Fibronectin (5 mM stock in PBS).
4. Phosphate buffer saline (PBS) pH 7.4.
5. Opti-MEM.
6. 70% ethanol.
7. Lipofectamine 2000 (Thermo Fisher Scientific) or other transfection reagent.
8. Complete Medium (CM): DMEM plus 10% heat-inactivated fetal bovine serum.
9. Live cell imaging solution (140 mM NaCl, 2.5 mM KCl, 1.8 mM $CaCl_2$, 1 mM $MgCl_2$ 20 mM HEPES pH 7.4 mOsm = 300) supplemented with 20 mM D-glucose (1 M stock, filter sterilized).
10. SNAP-Cell 647-SiR (New England Biolabs) and ATTO590-CA (*see* **Note 1**). Halo and SNAP dye ligands are suspended in DMSO (1 mM stock) and stored at −20 °C until immediately before labeling.
11. 0.02 µm diameter crimson fluorescent carboxylate-modified microspheres (For instructions on how to make a bead sample *see* **Note 2**).
12. CCL-2 HeLa cells or other cell model.
13. Bath sonicator.
14. STED microscope with 590 and 650 nm excitation sources and a ≅775 nm depletion laser.

3   Methods

3.1   Cleaning and Preparation of Imaging Dishes

Organic dyes tend to be sticky and cleaning the glass surface is necessary to prevent dyes from attaching to the cover glass, which can lead to high background (particularly for imaging of structures at the plasma membrane like endocytic events). Additionally we treat the glass surface with fibronectin to spread the cells so that they are flatter and geometrically more suitable for imaging.

1. Sonicate the glass bottom dishes for 15 min in 1 M KOH in a bath sonicator (*see* **Note 3**).

2. Carefully wash off the KOH by rinsing three times with ultra-pure water.

3. To sterilize the dishes, spray with 70% Ethanol in a cell culture hood and let them dry before proceeding to the next step.

4. Apply 1 mL of Fibronectin solution (5 μm/mL final concentration in PBS) to the dishes and incubate for 2 h at 37 °C. Wash three times with PBS.

3.2   Transfection and Labeling of Living Cells

In this protocol we will image SNAP-tagged transferrin receptor (TfR-SNAP) and Halo-tagged Clathrin Light Chain (Halo-CLC) as previously described [5] (*see* **Note 4**). SNAP tag fusions of your favorite putative endocytic markers/machinery components can be used instead of transferrin receptor. Alternatively, purified fluorescent ligands can be imaged. For example, fluorescent transferrin labeled with Alexa Fluor 594 is STED compatible and available to purchase from Thermo Fisher Scientific.

1. Seed 200,000 CCL-2 HeLa cells (*see* **Note 5**) on the KOH cleaned and fibronectin coated dishes prepared as described in Subheading 3.1.

2. The following day, transfect Halo-CLC and TfR-SNAP with your preferred method of transfection (*see* **Note 6**).

3. 16–20 h after transfection, label the transfected cells with ATTO590-CA and SNAP-Cell 647-SiR.

Dilute SNAP-Cell 647-SiR and ATTO590-CA into CM to a final concentration of 1 μM for each dye ligand in 150 μL of CM. Slowly aspirate old medium and apply 150 μL of the labeling solution to the center of the imaging dish (this works well with 14 mm glass diameter dishes. You may need to increase the volume of the labeling solution when using dishes with a bigger glass well). Incubate cells with substrates for 1 h at 37 °C/5% CO_2 in a standard mammalian cell culture incubator. Wash cells three times with CM and let excess dye washout for 1–3 h at 37 °C/5% CO_2 in a standard mammalian cell culture incubator. Gently aspirate CM. Wash three times with PBS. Add 2 mL of live cell imaging solution supplemented with 20 mM D-glucose to the dish.

3.3 Imaging Parameters

Clathrin and transferrin-fusion proteins are now labeled with the STED compatible dyes ATTO590 and SiR (Fig. 1). Any custom/commercial STED microscope with 590 and 650 nm excitation sources and a $\cong$775 nm depletion laser can be used to image this dye combination. STED imaging of living mammalian cells should be carried out at physiological temperature (37 °C) and, if possible, with CO_2 (infusion) to maintain healthy cells throughout imaging (*see* **Note 7**). We recommend equipping the microscope with either a heated box or a stage-top incubator. In the table below, we outline the settings used when imaging on either a custom built or a commercial Leica TCS SP8 STED 3× microscope. The same dye pair can also be used with any commercial/custom-built microscope with two pulsed excitation laser sources at 594 nm and 640 nm (including the Abberior 2C STED).

	Custom built	**Commercial Leica TCS SP8 STED 3×**
Excitation	594 nm and 650 nm pulsed diode excitation lasers	White light source. Select 594 nm and 650 nm as excitation wavelengths
STED laser	775 nm pulsed depletion laser	775 nm pulsed depletion laser
Scanning	16 kHz resonant scanner	8 kHz resonant scanner
Detection	APD1: 604–644 APD2: 665–705	HyD 1 set to 604–644 HyD 2 set to 665–705

Although commercial systems (like the Leica TCS SP8 STED 3X) are equipped with an auto-alignment routine, it is important to check that the auto alignment has worked correctly. We recommend doing so with 0.02 µm diameter crimson fluorescent carboxylate-modified microspheres (*see* **Note 2**).

Below we will outline important parameters to consider when imaging in order to achieve optimal resolution and reduce photobleaching:

1. *STED Delay control*: By adjusting this parameter, you vary the delay of the STED depletion laser pulse with respect to the excitation pulse. It is important to adjust the delay at the beginning of each imaging session to optimize resolution. While scanning

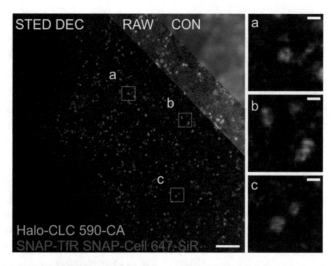

Fig. 1 Live-cell imaging of Halo-CLC and TfR-SNAP. First frame of a live-cell STED image sequence of HeLa cells expressing Halo-CLC (labeled with ATTO590-CA) and TfR-SNAP (labeled with SNAP-Cell 647-SiR). In this example a frame size of 1024 × 1024 pixel was chosen with 18.9 nm pixel size (*see* Subheading 3.2, **step 2**). Scale bars are 2 μm in the overview image and 200 nm in panels **a**, **b**, and **c**. *CON* confocal, *RAW* STED raw data, *STED* STED image deconvolved with python-microscopy (*see* Subheading 3.3)

with both excitation and STED laser on, find the delay position that maximizes STED depletion for both dyes (minimum fluorescence signal).

2. *Frame and Pixel size*: We recommend using a pixel size of 20 nm or smaller. There is a trade-off between pixel size, field of view size, and frame rate. We suggest using an image format of 512 × 512 pixels or 1024 × 1024 pixels when acquisition speed is not the limiting factor. For example, the live-cell sequence shown in Fig. 1 was acquired with one 2-color frame every ≅4 s (1024 × 1024 pixels image format, 18.9 nm pixel size, 8 kHz resonant scanner, and 16 line accumulations). Zooming in into a smaller area while keeping the same pixel size and changing the image format (for example 512 × 512) will allow you to go twice as fast if faster acquisition rate is necessary. Reducing the number of line accumulations or averages (in the case of samples with stronger signal) will also speed up the acquisition rate.

3. *Scanning speed*: For live-cell samples, we recommend using the fast resonant scanner. Fast beam scanning reduces pixel dwell time and photobleaching.

4. *Scanning mode*: Since the emission spectra of ATTO590 and SiR are very close, we recommend using the line-sequential acquisition to minimize cross-talk between channels.

5. *STED power*: For live samples, we recommend using the minimal STED laser power needed to achieve your desired resolution while minimizing photobleaching and phototoxicity. For Leica microscopes equipped with a Onefive Katana HP laser, we use 40–50% STED power (110–140 mW, measured at the sample) when the laser is powered to its maximum.

6. *Image processing and deconvolution*: We suggest smoothing the raw STED data using Image J [6] for presentation purposes (use a Gaussian filter with 1 pixel standard deviation). To reduce noise, movies are deconvolved using the deconvolution algorithm of the python-microscopy package available at http://python-microscopy.org.

4 Notes

1. SNAP-Cell 647-SiR is available for purchase from New England Biolabs. While a wide variety of Halo tag substrates are available from Promega, ATTO590-CA was produced in the lab [5]. ATTO590-CA can be made in the lab following the protocol provided by Promega at https://www.promega.com/resources/protocols/technical-manuals/0/Halotag-technology-focus-on-imaging-protocol/. For this purpose we recommend using The HaloTag® Amine (02) Ligand (Promega) and ATTO590 NHS ester (Sigma).

2. To make a bead sample, apply 150 μL of 0.01% poly-L-lysine to the center of a clean cover glass. Let stand for 5 min. Resuspend your stock solution of beads by vortexing briefly. Dilute beads with PBS and place in bath sonicator for 5 min to separate aggregated beads. We recommend starting with a 1:10,000 dilution and adjusting as needed. Rinse the cover glass three times with PBS and then apply 150 μL of diluted beads to the glass. Wait 10 min for beads to settle onto the glass. Rinse cover glass three times with fresh PBS. Invert glass + beads onto microscope slide with mounting medium. For best results, allow the mounting medium to polymerize before imaging. We suggest using crimson FluoSpheres (ThermoFisher) as they can be excited by both 594 nm and 650 nm excitation lines.

3. The lab is equipped with a Branson 2800 ultrasonic cleaner but any equivalent cleaner will work.

4. To generate Halo-CLC, Halo tag was fused to the N-terminus of CLC as described in [5] for SNAP-CLC. To generate

TfR-SNAP, SNAP was fused to the C-terminus of transferrin receptor.

5. Any transfectable adherent cell line of your choice can be used instead of HeLa cells.

6. For Lipofectamine 2000 transfection, we suggest complexing 0.5 μg of Halo-CLC and 1 μg of TfR-SNAP with 10 μL of Lipofectamine in a total volume of 500 μL of Opti-MEM. Incubate for 5 min at RT and apply the complexes to the cells in 2 mL of complete medium. We also successfully transfect HeLa cells with FuGENE HD (Promega) and electroporation.

7. When imaging without CO_2 infusion we recommend limiting the imaging time to 2 h.

Acknowledgments

We thank Stephanie Wood Baguley for generating the plasmids used in this protocol. We thank Dr. Joerg Bewersdorf, Dr. Manuel Jütte, Dr. Edward Allgeyer, and Mark Lessard for critically reading the manuscript. This project was supported by the Wellcome Trust (095927/A/11/Z, 092096) and the G. Harold & Leila Y. Mathers Charitable Foundation.

References

1. Hell SW, Wichmann J (1994) Breaking the diffraction resolution limit by stimulated emission: stimulated-emission-depletion fluorescence microscopy. Opt Lett 19(11):780–782
2. Schermelleh L, Heintzmann R, Leonhardt H (2010) A guide to super-resolution fluorescence microscopy. J Cell Biol 190(2):165–175
3. Tonnesen J, Nadrigny F, Willig KI, Wedlich-Soldner R, Nagerl UV (2011) Two-color STED microscopy of living synapses using a single laser-beam pair. Biophys J 101(10):2545–2552
4. Lukinavicius G, Umezawa K, Olivier N, Honigmann A, Yang G, Plass T, Mueller V, Reymond L, Correa IR Jr, Luo ZG, Schultz C, Lemke EA, Heppenstall P, Eggeling C, Manley S, Johnsson K (2013) A near-infrared fluorophore for live-cell super-resolution microscopy of cellular proteins. Nat Chem 5(2):132–139
5. Bottanelli F, Kromann EB, Allgeyer ES, Erdmann RS, Wood Baguley S, Sirinakis G, Schepartz A, Baddeley D, Toomre DK, Rothman JE, Bewersdorf J (2016) Two-colour live-cell nanoscale imaging of intracellular targets. Nat Commun 7:10778
6. Abramoff MD, Magalhaes PJ, Ram SJ (2004) Image processing with ImageJ. Biophoton Int 11(7):36–42

Chapter 15

Measuring Clathrin-Coated Vesicle Formation with Single-Molecule Resolution

François Aguet and Emanuele Cocucci

Abstract

High-resolution fluorescence microscopy is increasingly contributing to our understanding of molecular processes. By utilizing single-molecule intensity information, imaging experiments can be rendered quantitative, yielding insights into the stoichiometry and kinetics of the components of a molecular assembly. Here, we describe the experimental and analytical steps needed to study the assembly of clathrin-coated vesicles with single-molecule resolution, using total internal reflection fluorescence microscopy. Many components of the protocol are broadly applicable to the characterization of other molecular processes.

Key words Imaging, Single molecule, TIRF, Clathrin-coated vesicles

1 Introduction

The formation of vesicular carriers by the sequential assembly of coatomers is a highly conserved process in eukaryotes [1]. It ensures sorting and transfer of specific lipids and proteins across the organelles which comprise a cell [2]. Among different carriers, those responsible for trafficking components from the plasma membrane to the early endosomes are enclosed by a 180-KDa protein named clathrin [3]. Clathrin-coated vesicles have a primary role in the internalization of receptor-bound ligands and of extracellular fluids [4], and are also implicated in the recycling of secretory vesicle components in neuroendocrine cells [5, 6]. The assembly of clathrin-coated vesicles occurs by the sequential recruitment of their basic units, the clathrin triskelion, and the adaptor protein complex, AP2. These two components are essential for life [7] and for the formation of clathrin-coated vesicles, which also requires additional accessory proteins [8].

The development of fluorescent protein chimera [9] and high-resolution fluorescence microscopy made it possible to study the

The original version of this chapter was revised. A correction to this chapter can be found at
https://doi.org/10.1007/978-1-4939-8719-1_19

Laura E. Swan (ed.), *Clathrin-Mediated Endocytosis: Methods and Protocols*, Methods in Molecular Biology, vol. 1847,
https://doi.org/10.1007/978-1-4939-8719-1_15, © Springer Science+Business Media, LLC, part of Springer Nature 2018

assembly of clathrin-coated vesicles in living cells [10]. In a typical imaging time-series, clathrin-coated vesicle assembly events occur at high spatial density and emit a noisy fluorescence signal, requiring sensitive and accurate analytical tools for the detection and tracking of all observed events [11, 12]. The development of such tools enabled the identification of the general stages of clathrin assembly and led to the hypothesis of discrete assembly checkpoints, regulated by cargo and dynamin recruitment [11, 13, 14]. Importantly, unbiased automated analyses are crucial for capturing the full diversity of molecular events, including rare but significant events [15] such as clathrin-coated vesicle formation in partially dynamin-depleted cells [16]. Due to several factors, including a limited photon budget and the large number of proteins constituting individual coated pits, it is in general not yet feasible to track individual proteins over the course of the entire assembly process. However, using single-molecule intensity calibration data and statistical modeling, it is possible to resolve the arrival of individual proteins during the initial stages of coated pit formation, which is initiated by stochastic interactions between clathrin and AP2 [17].

In this chapter, we describe an approach to study clathrin-coated vesicle formation with single-molecule resolution. To this end, we rely on an imaging setup based on total internal reflection microscopy (TIRF-M), which excites fluorescent molecules in close proximity (~100–200 nm) to the glass surface, resulting in high signal-to-noise ratios and enabling the detection of single molecules [18]. We detail the steps involved in the calibration of the system, in the preparation of coverslips, in the mounting and imaging of cells, and finally in the analysis of the resulting data. As part of the latter step, we describe an approach to identify the number of adaptor protein complexes during the first phase of clathrin-coated pit formation in experimental conditions where a fraction of unlabeled adaptor complexes are present [17]. Many components of the methods described herein can be applied to the study of other molecular processes amenable to high-resolution imaging, in particular processes localized at the plasma membrane, such as virus budding or the clustering of receptors.

2 Materials

2.1 Equipment

Store at room temperature, unless stated otherwise.

1. *Glass coverslips.* # 1.5 round 25 mm glass coverslips.

2. *Total internal reflection microscopy system.* The imaging setup described here is based on a Marianas system (Intelligent Imaging Innovations Inc., CO); any equivalent setup is suitable for the experiments described in this chapter (*see* **Note 1**). The system is composed of an Axiovert 200 M microscope

equipped with an Alpha Plan-Apo 100× objective (1.46 NA, Carl Zeiss, DE) and a TIRF slider with manual angle and focus controls (Carl Zeiss, DE). The microscope stage, completely enclosed in a temperature-controlled chamber, is equipped with a Peltier temperature-controlled sample holder kept at 37 °C (20/20 Technology, Wilmington, NC). The illumination is supplied by a solid-state laser (488 nm wavelength, Sapphire, 50 mW, Coherent Inc., CA). The excitation light is coupled through an acousto-optical tunable filter into a single mode optical fiber, carried to the TIRF slider and reflected into the objective using a multiband dichroic mirror (Di01-R405/488/561/635, Semrock, Rochester, NY). The emission light is collected by the objective, passes the dichroic mirror and a multiband emission filter (we use FF01-466/523/600/677, Semrock, Rochester, NY), and is projected onto an electron multiplying charge-coupled device (EMCCD) camera (QuantEM, Photometrics). A 2× magnification lens placed in front of the EMCCD camera is employed to reach a final pixel size of 80 nm. Microscope operation and image acquisition are controlled by the SlideBook software (Intelligent Imaging Innovations Inc., CO).

3. *Humidified chamber for coverslip incubation.* Any box that generates a humid environment while keeping coverslips from direct contact with water is suitable. We suggest an empty pipette tip box designed for storing 200 μL tips. These boxes come with a lid and have a perforated surface (originally holding the tips) to place the coverslips on. Fill the box with ~15 mL of water; the holes will provide sufficient humidification once the lid is closed.

4. *Fluorescent beads.* 0.1 μm carboxylated beads (Ex/Em: 505/515 nm; Life Technologies Corporation, NY) stored at 4 °C.

5. *NHS-LC-Biotin.* Amine reactive biotin, stored desiccated at −20 °C.

6. *Coverslip rack.* Teflon rack to vertically hold glass coverslips during cleaning procedures.

7. *Holder for imaging.* A coverslip holder that can fit the microscope is required. Several solutions are available, we adopted the Attofluor Cell Chamber (Life Technologies Corporation, NY).

8. *Dialysis unit.* Mini Dialysis unit (molecular weight cut off: 10^4 Daltons).

9. *Bottletop filter unit.* Reusable bottletop filter unit with removable membrane of 0.2 μm pore size for vacuum filtration.

10. *Protease inhibitors.* Protease inhibitors are necessary to preserve protein stability upon cell lysis. We used protease inhibitor cocktail tablets (Pierce, IL).

11. *Sonicator.* A benchtop water bath sonicator (we use one from Branson-Emerson, MO).

12. *Glow discharge unit.* Glow discharge units charge negatively or positively metal or glass surfaces, facilitating the spreading of aqueous solutions (e.g., Pelco EasyGlow Discharge Cleaning System, Ted Pella, CA).

13. *Desiccator.* A desiccator vacuum-based unit with a diameter of ~200 mm.

2.2 Solvents, Solutions, and Cell Culture Reagents

Prepare all solutions with ultrapure water (purified deionized water to attain a sensitivity of 18 MΩ cm at 25 °C).

1. *Water.* Filter 1 L of ultrapure water in a clean bottle using a vacuum-based bottletop filter unit (*see* **Note 2**); fill a clean sprinkle with 100 mL to be used during the coverslip preparation.

2. *Ethanol.* Filter 1 L of denatured ethanol into a bottle using a vacuum-based bottletop filter unit (*see* **Note 2**). Fill a glass jar with the filtered ethanol such that it will entirely cover a rack loaded with glass coverslips.

3. *1 M NaOH.* In a 100 mL cylinder poor 50 mL of water, dissolve 3.99 g of NaOH. Then bring to final volume of 100 mL with water, and store in a bottle at 4 °C.

4. *0.5 M EGTA.* In a 100 mL beaker pour 50 mL of water and add under constant stirring 19 g of EGTA. Warm the solution and increase pH with 1 M NaOH to facilitate dissolution of the crystals.

5. *1 M HEPES.* In a 100 mL cylinder poor 50 mL of water, dissolve 23.8 g of HEPES under stirring. Bring the solution to 90 mL with water and buffer to pH 7.5 using NaOH, then bring to final volume of 100 mL with water, and store in a bottle at 4 °C.

6. *20% Triton X-100.* in a 250 mL beaker pour 20 mL of Triton X-100 and 80 mL of water, mix by continuous stirring; transfer to a bottle.

7. *Phosphate Buffer Saline (PBS).* 137 mM NaCl, 2.7 mM KCl, 10 mM Na_2HPO_4, 1.8 mM KH_2PO_4. In a graduated 1 L cylinder or glass beaker pour 500 mL of water. Under constant stirring, add salts (NaCl 8 g; KCl 0.2 g; Na_2HPO_4 1.44 g; KH_2PO_4 0.24 g). Bring the volume to 900 mL and adjust pH to 7.3 with HCl, then bring to final volume of 1 L. Filter in a clean 1 L bottle using a bottletop filter unit (*see* **Note 2**). Load a sprinkle bottle with 100 mL of PBS to be used during coverslip preparation.

8. *Cell lysis buffer.* 50 mM HEPES, pH 7.5, 150 mM NaCl, 15 mM MgCl$_2$, 1 mM EGTA, 10% glycerol, and 1% Triton X-100. In a 250 mL beaker, pour 50 mL of water and under constant stirring add 876 mg of NaCl, 142 mg of MgCl$_2$, 5 mL of HEPES 1 M, 200 mL of 0.5 M EGTA, 10 mL of glycerol, and 5 mL of 20% Triton X-100.

9. *PLL-g-PEG solution.* The copolymer composed of poly-L-lysine (PLL) and poly(ethylene glycol) (PEG) with 20% of biotinylated molecules (PLL(20)-g[3.4]-PEG(2)/PEG(3.4)-biotin [20%]) is dissolved in filtered PBS at 1 mg/mL in water and stored in aliquots of 200 μL at −20 °C.

10. *Streptavidin.* Dissolve streptavidin in PBS at 1 mg/mL, store in aliquots of 100 μL at −20 °C.

11. *1 mg/mL Fibronectin* from bovine plasma. Store at 4 °C.

12. *Biotinylated fibronectin solution.* Dissolve in 1.5 mL tube 1 mg of NHS-LC-Biotin in 1 mL of PBS to obtain ~2 mM solution (*see* **Note 3**). To the 1 mL of 1 mg/mL fibronectin solution (~4.5 μM), add 25 μL of NHS-LC-Biotin solution to reach 10× excess of NHS-LC-Biotin. Incubate at room temperature for 1 h. During this time, gently mix the solution by tipping the vial. Load a 10 KDa mwCO dialysis unit with the solution and put it in a 5 L plastic beaker filled with 3 L of PBS for 3 h. Exchange the PBS and let dialyze overnight at *4 °C (see* **Note 4***)*.

    Recover the dialyzed solution, add glycerol to 20%, and mix thoroughly. Aliquot the biotinylated-fibronectin solution in 50 μL volumes and store at −80 °C (*see* **Note 5**).

13. *Fluorescent beads in PBS.* Dilute fluorescent beads in PBS to obtain ~4 × 10^7 beads in 100 μL, which corresponds to a 10^5-fold dilution. Thus, prepare two consecutive dilutions, in the same buffer, first 1:1000 to obtain a 1 mL stock solution then 1:100 to obtain the working solution. Store all bead solutions at 4 °C or on ice during the experiment.

14. *Cell media preparation.* Under a cell culture hood, in a clean and sterilized 500 mL bottle, filter using a sterile bottletop filter 445 mL DMEM, 50 mL fetal bovine serum (final concentration 10%) and 5 mL of a sterile solution containing 10,000 units/mL penicillin, 10,000 mg/mL of streptomycin (*see* **Note 6**). Store at 4 °C.

15. *Imaging media.* Aliquot MEMα in 50 mL conical tubes; store at 4 °C. 2 h prior to imaging, add to 1% FBS to MEMα and warm up to 37 °C (*see* **Note 7**).

16. *Cell line for imaging.* We use BSC1 cells stably overexpressing AP2S1-EGFP.

2.3 Software

For the data acquisition and analytical parts of this protocol, a workstation or personal computer with the following software installed is required:

- Image acquisition software. Many acquisition software solutions exist; commercial imaging systems are often paired with dedicated software. The analytical components of this protocol assume that image data are exported in multipage TIFF format (also called TIFF stacks).

- Matlab (version 2011a or above).

- cmeAnalysis package for Matlab installed (available from https://www.utsouthwestern.edu/labs/danuser/software/#cme-anchor).

3 Methods

3.1 Cell Lysate Preparation

Cell lysate obtained from fluorescent protein-expressing cells is used as source of single fluorophores and is used to calibrate the imaging system (*see* **Note 8**).

1. Grow the cells in their appropriate medium (*see* **Note 9**).

2. Once a 10 cm diameter petri dish is confluent (~4 × 10⁶ cells), remove (in a cell culture hood) the medium and rinse with sterile PBS.

3. After removing PBS by suction, add 1 mL of trypsin cell culture grade solution and place at 37 °C for few minutes.

4. Resuspend the cells in the trypsin by gently pipetting and put the cell suspension in a 15 mL conical tube.

5. Add 5 mL of sterile PBS and centrifuge for 5 min at $800 \times g$ at 4 °C.

6. Remove the supernatant and suspend the cells in 5 mL of sterile PBS.

7. Repeat **steps 5** and **6** twice. Then resuspend the cells in 1 mL of sterile PBS and place in a 1.5 mL tube.

8. Move the cells onto ice. All the subsequent steps should be carried out at 4 °C or on ice to prevent protein degradation.

9. Centrifuge at 4 °C for 5 min at $800 \times g$.

10. Gently resuspend the cells in 100 µL of 4 °C lysis buffer supplemented with proteinase inhibitor cocktail. Leave on ice 10 min.

11. Centrifuge the lysed cells at $800 \times g$ 5 min at 4 °C to discard nuclei. Collect the supernatant and store in 10 µL aliquots at −80 °C. Use one aliquot to measure the protein concentration, which is expected in the range of ~7 mg/mL of proteins.

3.2 Clean Glass Coverslip Preparation

1. Place glass coverslips in a Teflon rack holding them vertically.

2. Prepare three jars containing ~200 mL of filtered water, ethanol, and NaOH. The jars should contain the coverslips loaded rack and the solutions should completely cover the coverslips leaving adequate room at the top to avoid spilling (1 cm).

3. Sonicate the coverslips in filtered ethanol for 30 min in a water bath sonicator.

4. Remove the coverslips from filtered ethanol.

5. Adsorb the solution by placing the coverslip-loaded rack on laboratory tissues.

6. Rinse the coverslips in ultrapure filtered water by dipping the rack in the jar for 3 min and pulling them out briefly (5 s) five times to efficiently exchange the water on the glass surface. Do not let the coverslips dry when pulled out.

7. Place the coverslips into 1 M NaOH and sonicate for 30 min (*see* **Note 10**).

8. Repeat rinse **step 6**.

9. Sonicate the coverslips into filtered ethanol for 30 min.

10. Remove the excess of ethanol by tapping the rack on laboratory tissue and place the rack with the coverslips in an oven at 120 °C to dry.

11. Expose the coverslips to air plasma inside a glow discharge unit operated at 50 mA for 3 min (*see* **Note 11**).

Glass coverslips processed as described will be referred as *clean glass coverslips* and are directly used in the coating of clean coverslips with PLL-g-PEG mixture protocol (Subheading 3.3) and to image beads or single fluorescent molecules (Subheading 3.5). *Clean glass coverslips* cannot be stored since the air plasma charge lasts only for a few minutes. If some coverslips are not used right away, we suggest storing them in an oven at 120 °C. Another cycle of sonication in ethanol and exposure to air plasma should precede any further step in the protocol.

3.3 Coating of Clean Glass Coverslips with PLL-g-PEG Mixture

1. Immediately after obtaining *clean* glass coverslips (Subheading 3.2), sit a coverslip in the humidified chamber.

2. Add 50 μL PLL-g-PEG solution on the center of the coverslip.

3. Place a second coverslip on the top of the first one. Let the "sandwiched" glass coverslips incubate for 1 h.
   Repeat **1–3 steps** four times or until sufficient coverslips have been prepared (*see* **Note 12**).

4. Rinse the functionalized surfaces with filtered water one by one (*see* **Note 13**).

5. Store the dried coverslip under reduced pressure in a desiccator at room temperature. Under those conditions the coverslips can be stored up to 2 weeks [19].

3.4 Cell Plating on Fibronectin-Coated PLL-g-PEG Glass Coverslips

1. Place a glass coverslip (prepared in Subheading 3.3) in the humidified chamber with the PLL-g-PEG coated surface facing up. Dilute streptavidin solution (Subheading 2.2, **item 6**) to the working concentration of 0.2 mg/mL in PBS with 0.02% Tween 20 (*see* **Note 14**) and place 100 μL on the glass coverslip. Incubate for 30 min at room temperature.

2. Rinse the coverslip with filtered PBS and place it in the humidified chamber (*see* **Note 13**).

3. Add 100 μL of 5 nM biotinylated fibronectin diluted in PBS with 0.025% Tween 20 (*see* **Note 14**), and incubate for 30 min at room temperature.

4. Place a Teflon rack in a PBS-filled beaker. Rinse the coverslips with PBS (*see* **Note 13**) and place them on the rack.

5. Under cell culture hood, place the coverslips into 6-well dishes (*see* **Note 15**). Wash twice with sterile PBS and one with cell medium.

6. Suspend the cells by trypsinization, wash the cells in PBS and resuspend them in complete medium at 25×10^4 mL.

7. Use suction to remove medium from the six wells (**step 5**) and plate ~50×10^4 cell on the coverslips by adding 2 mL of resuspended cells in the six well plate.

8. Different cell types spread on the glass coverslips in variable amount of time. Check the cells every hour until cells are completely spread (*see* **Note 16**).

3.5 System Calibration by Imaging Fluorescent Beads

1. Place a *clean glass coverslip* (prepared in Subheading 3.2) in the humidified chamber.

2. Dilute 1:1 fluorescent beads working solution in PBS and spread 100 μL on the glass coverslip. This concentration permits to homogenously sample the field of view (*see* **Note 17**).

3. Let the beads adsorb for 5 min.

4. Remove the majority of the bead solution by dabbing the edges of the coverslip onto a laboratory tissue.

5. Insert the glass coverslip into a holder, cover it with 1 mL of PBS and place it on the microscope stage.

6. Find the focal plane by focusing on the beads.

7. Choose a field of view where the beads are homogenously distributed over the entire field of view and evenly spaced (*see* **Note 18**). If no such area can be found, adjust the bead concentration as needed and prepare a new coverslip.

8. Acquire single plane images, varying exposure time (e.g., 25, 50, 100, 200, and 500 ms), laser power (e.g., 0.5, 2, 5, 10, 15, 20 mW; *see* Fig. 2), and camera multiplication gain. Acquire images at 3–5 different values per variable while leaving the others fixed to validate the linearity of the system. The acquired images will also be used to define the homogeneity of TIRF excitation intensity and the homogeneity of the field of view (*see* **Note 19**).

9. Analyze the obtained data (Subheading 3.7; *see* **Note 20**).

3.6 Single Molecule Calibration

1. Place a *clean* glass coverslip (prepared in Subheading 3.2) in the humidified chamber.

2. Dilute the cell lysate (prepared in Subheading 3.1) containing the fluorescent proteins 1:1000 in PBS. In 100 μL of the diluted solution, add 1–10 μL of fluorescent beads working solution (Subheading 2.1, **item 5**) to the diluted cell lysate (*see* **Note 21**).

3. Spread the bead-lysate mix from **step 2** (Subheading 3.6) on one clean glass coverslip (prepared in Subheading 3.3).

4. Remove the majority of the beads solution by dabbing the edge of the coverslip onto a laboratory tissue.

5. Insert the glass coverslip in a holder and place it on the microscope stage.

6. Find the focal plane by focusing on the beads (*see* **Note 22**). From here on, all the imaging operations are performed by acquiring single images to avoid fluorescent protein photobleaching by continuous imaging.

7. Slightly move field of view, and acquire an image to verify that no beads are present (*see* **Note 23**).

8. Increase the power to visualize single molecules; acquire one image to verify that the field of view is in focus and that the fluorescent proteins are evenly scattered. If no such area can be found, adjust the bead concentration as needed and prepare a new coverslip (*see* **Note 24**).

9. Acquire time series of fluorescent proteins as follows: 200–300 consecutive frames (with no intervals), exposure time: 30, 50, 100, and 200 ms. The excitation shutter should be closed in between frames. Constant imaging of the same area will induce photobleaching of single fluorophores. Acquire at least 5 fields of view per exposure condition. Over the course of each acquisition, you should observe instantaneous loss of fluorescence

signal, corresponding to single-step photobleaching, at the location of individual molecules.

10. Analyze the obtained data (*see* Subheading 3.8).

3.7 Live Cell Imaging

Live cell imaging should follow the calibration experiments described in Subheadings 3.5 and 3.6; to ensure the highest accuracy, the single molecule calibration (Subheading 3.6) should be performed before every experiment.

1. Place a coverslip with cells (prepared in Subheading 3.4) in the holder for imaging.

2. Rinse the coverslip with 1 mL PBS.

3. Remove PBS and add 1 mL of imaging media previously warmed at 37 °C.

4. Place the holder for imaging in the microscope (*see* **Note 25**).

5. To avoid z drifting, wait ~10 min to allow the sample to equilibrate to the temperature of the microscope.

6. Locate a cell of interest (*see* **Note 26**). Well-spread cells expressing AP2S1-EGFP have homogenous cytosolic intensity and clathrin-coated vesicles appear as bright diffraction-limited spots.

7. Acquire a time series at ~5 Hz or higher to visualize the first recruitment steps of AP2s at the plasma membrane, during clathrin coated pit initiation (*see* **Note 27**).

3.8 Analysis of Calibration Data

All image analyses are described for Matlab (MathWorks, Natick, MA). Note that for all Matlab functions mentioned in this protocol, detailed documentation is available via built-in documentation (through the help/doc functions).

1. The first step of the calibration consists in estimating the parameters of a Gaussian model of the point spread function (PSF), which will be used to detect clathrin-coated vesicles. Begin by loading the bead images (acquired in Subheading 3.5) into Matlab. Then, use the pointSourceDetection function to detect the beads and estimate the parameters of the Gaussian model for each bead (use mode "xyasc" to estimate the position, amplitude, standard deviation, and local background for each bead). For the numerical fit, an initialization value of 1.3–1.5 (pixels) for the standard deviation is usually adequate, and should be robust for the experimental setup described here (otherwise, slightly larger values up to ~2 are recommended). The distribution of the fitted standard deviations should be approximately normal, and the calibration value for the PSF model can be calculated by taking the median value (*see* Fig. 1).

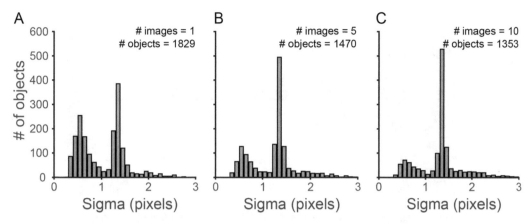

Fig. 1 Calibration of the point spread function model. Fluorescent beads were adsorbed to a glass coverslip. The same field was imaged by TIRF-M for ten consecutive frames (1 ms exposure, 19 mW power). A Gaussian point spread function model was then fitted to a single frame (**a**), an average of 5 frames (**b**), and an average of ten frames (**c**), with the standard deviation σ (sigma) as a free parameter. The resulting distributions of σ show a decrease in spurious background detections ($\sigma < 1$) with an increased number of averaged frames. For the imaging setup described here, the optimal value of σ is 1.35 pixels for sources with a peak emission wavelength at ~510 nm (for a pixel size of 80 nm, $\sigma = 1.35$ corresponds to ~108 nm)

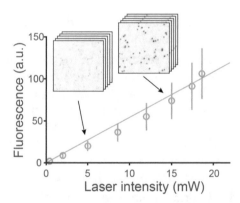

Fig. 2 Linearity of fluorescence emission as a function of laser excitation power. Fluorescent beads were adsorbed to a glass coverslip. The same field was imaged at eight different laser powers (0.5, 2, 5, 9, 12, 15, 17, and 19 mW, respectively and 1 ms exposure time) for 10 consecutive frames each. Detection of fluorescent objects using $\sigma = 1.35$ (*see* Fig. 1) was employed to estimate the intensity of all beads detected at each laser power setting. Average bead intensities ± s.d. are shown in orange; the linear regression is shown in blue ($r = 0.98$)

2. The second step of the calibration consists in verifying the linearity of the imaging system. Using laser intensity and/or exposure time calibration image series, run pointSourceDetection on the image of the series with the highest intensity or exposure to detect the beads (use mode "xyac," and provide the standard deviation calculated in the previous step as input). Use the resulting positions to run pointSourceDetection on the remaining images of the calibration series. Finally, plot the estimated amplitudes for each condition against the corresponding power

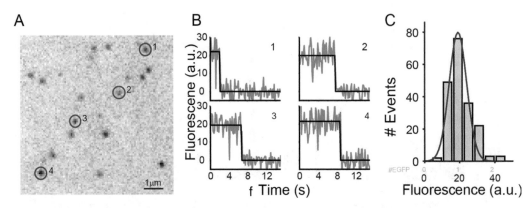

Fig. 3 Single EGFP fluorescence intensity calibration. EGFP molecules were adsorbed to glass coverslips by electrostatic binding and imaged with TIRF (100 ms exposure, 19 mW power, 100 consecutive frames). (**a**) Cropped region of a single frame, showing single EGFPs (inverted contrast). (**b**) Fluorescence intensity traces corresponding to the molecules highlighted in (**a**). The background-corrected fluorescence intensities (blue) show single-step photobleaching of EGFP. A step fitting function (black) was applied to the intensity traces to determine single-molecule intensities. (**c**) Single EGFP intensities follow a Gaussian distribution. For the data shown, the intensity was estimated as 19 ± 5 (mean ± s.d.) from the measurement of 191 diffraction-limited spots derived from three different acquisitions

or exposure value, and verify the linearity of the relationship by calculating a linear regression (*see* Fig. 2).

3. Next, the parameters of the single-molecule intensity distribution are estimated, using the images acquired in Subheading 3.6. Since the single-molecule positions are fixed, a robust estimate of the positions can be obtained by first running the detection on an average intensity projection of the images (*see* **Note 28**). Using these position estimates, intensity traces for each molecule across the entire acquisition are obtained by running pointSourceDetection with the positions provided as inputs.

4. Plotting the intensity traces should reveal single-step photobleaching induced by continuous imaging; in general, traces with a single step correspond to a single fluorophore and the magnitude of the step to its intensity. The steps can be identified using a step fitting algorithm [20–24], and the single molecule intensity can be estimated from the resulting distribution of step magnitudes (*see* **Note 29**). The step sizes should be approximately Gaussian-distributed. Consequently, the single molecule intensity can be calculated from a Gaussian fit, yielding the mean intensity and its variance (Fig. 3). If multiple exposure conditions are used, repeat Subheading 3.7, **steps 3** and **4** for each condition, and verify that the mean intensities calculated with the Gaussian fit are a linear function of exposure time.

3.9 Analysis of Live-Cell Imaging Data

1. Detect and track the recruitment of AP2-EGFP in assembling clathrin-coated vesicles by using cmeAnalysis [11] (which utilizes the tracking framework of uTrack, used in [16, 17]).

Input the standard deviation of the PSF model (σ) as estimated in Subheading 3.7, **step 1** to ensure optimal detection of coated pits. Tracking parameters may need to be adjusted based on the experimental conditions. The parameters are defined in cost functions and summarized in Table 1; the values listed were used to analyze the data from [17]. Depending on the frame rate of the acquisitions, the parameters of the tracking function should be adjusted to guarantee optimal linking of individual detections between the frames (*see* **Note 30**). For more information on these parameters, refer to the Supplementary Materials of Jaqaman et al. [12].

2. Following the tracking step, filter individual traces to select those that: (1) show an increase in fluorescence intensity over time starting from the background intensity (indicative of an initiating and assembling pit); (2) do no collide with other objects (these trajectories are marked as composite tracks by the software and are excluded to avoid any ambiguity in resolving individual tracks); (3) appear after the time series started, and last longer than 15 s.

3. Apply a step fitting algorithm (20–24) to the traces selected in the previous step to identify stepwise intensity increases. At a 5 Hz acquisition frequency, the increments should be distinguishable by eye. We recommend inspecting and carefully verifying the step fitting results to screen for potential artifacts.

4. Collect the intensities corresponding to the first and second steps of each trace into two separate arrays (*see* **Note 31**).

5. Because the intensity of single fluorescent proteins follows a Gaussian distribution, the distribution of step intensities behaves as a Gaussian mixture, corresponding to the number of AP2-EGFPs recruited at each step. A cumulative distribution of the step intensities can be fitted using a Gaussian mixture model, assuming that all recruited molecules are fluorescently labeled. In cases where a fraction of the protein of interest remains unlabeled, it is necessary to account for the unlabeled fraction by incorporating a binomial distribution representing the labeled and unlabeled populations into the Gaussian mixture model. This model is formulated as follows:

$$f[k] = \sum_{i=1}^{M} a_i \cdot \binom{n_i}{k} \cdot q^k \cdot (1-q)^{n_i-k}$$

$$F(x) = \sum_{n=1}^{n_{\max}} f[n] \cdot \frac{1}{2}\left[1 + erf\left(\frac{x-(\mu \cdot n)}{\sigma \cdot \sqrt{2 \cdot n}}\right)\right]$$

where $f[k]$ is the distribution of fluorophores corresponding to a combination of M possible configurations, n_i is the number of proteins (labeled and unlabeled) corresponding to each configura-

Table 1
Tracking parameters

GapClose Param	
timeWindow	3–4 or larger
mergeSplit	1
minTrackLen	1
diagnostics	0
costMatrices(1)	
funcName	costMatLinearMotionLink2
linearMotion	0
minSearchRadius	3
maxSearchRadius	5
brownStdMult	3
useLocalDensity	1
nnWindow	timeWindow
kalmanInitParam	[]
diagnostics	[]
costMatrices(2)	
funcName	costMatLinearMotionCloseGaps2
linearMotion	0
minSearchRadius	3
maxSearchRadius	5
brownStdMult	3*timeWindow
linScaling	[1 0.01]
timeReachConfL	timeWindow
maxAngleVV	[]
gapPenalty	[]
resLimit	[]

tion, α_i is the relative proportion of each configuration (constrained to $\sum_i \alpha_i = 1$), and q is the probability of incorporating a labeled protein. $F(x)$ is the cumulative distribution of the Gaussian mixture model, where n_{max} is the largest number of proteins recruited in any of the configurations, and μ, σ are the mean and standard deviation of the single molecule intensity, respectively. The number of fluorescent proteins detected is a function of the number of molecules simultaneously recruited during each step, and this may be a combination of molecules (e.g., a majority of two AP2-EGFPs combined with a smaller fraction of four AP2-EGFPs, corresponding to $n_1 = 2$, $n_2 = 4$, $M = 2$; in our experiments the labeled fraction corresponded to $q{\sim}0.8$). Putative combinations must be explicitly tested by fitting their corresponding model to the intensity distribution of each step; the best fitting model is then identified by means of a model selection criterion (e.g., the Bayesian Information Criterion [25]. A model selection criterion is essential to avoid overfitting). Fit this model to the two arrays generated during the previous step, over a range of appropriate configuration, and select the best fitting model.

If multiple consecutive stages of a process are analyzed (e.g., two consecutive steps of AP2 incorporation), performing the fit jointly for all measurements is preferable, since the incorporation rate q is estimated only once, leading to more robust estimates and model selection.

4   Notes

1. Several TIRF imaging systems are commercially available. The following properties are essential for the experiments described in this chapter:

   - A temperature and humidity-controlled chamber is required for live cell imaging.

   - The sample should be imaged with a 488 nm laser at a power of ~5 mW and an exposure of 50 ms to ensure efficient detection of single EGFP molecules in TIRF (calibrated to a penetration depth of ~150 nm).

   - The pixel size, which corresponds to the sampling interval, should be smaller than the Nyquist limit, defined as $\lambda/(4*NA)$, where λ is the emission wavelength and NA the numerical aperture. In a system with an objective of $NA = 1.47$, the pixel size should therefore be smaller than ~85 nm for imaging EGFP.

2. Since the solution does not need to be sterile, reusable vacuum bottletop filters with 0.2 μm pores may be used.

3. NHS-LC-Biotin is extremely sensitive to humidity. Carefully bring the container to room temperature before weighting 1 mg of the compound, which is then transferred to the 1.5 mL tube. Subsequently, add the PBS and carefully observe the biotin dissolving by gentle pipetting.

4. To obtain efficient dialysis, keep the external solution under constant slow stirring.

5. Instantly freezing the protein solution in liquid nitrogen or a mixture of dry ice and ethanol improves its conservation.

6. The cell media recipe described here is optimized for culturing the BSC1 green monkey cells used in this protocol. Other cell lines may require different culturing conditions.

7. Adding 1% of FBS to the imaging media improves cell viability without significantly increasing background. If administration of drugs is part of your experimental design, remember that the albumin present might largely adsorb the small molecules, affecting experiment outcome.

8. If the fluorescent protein is part of a chimeric protein that forms or integrates into stable complexes, stoichiometry is an important factor for quantitative measurements. For clathrin-coated vesicles, if EGFP-tagged clathrin light chain-A is ectopically expressed, it will compete for binding to clathrin heavy chain with endogenous light chain A/B. Therefore, the assembled triskelia will have 0–3 bound fluorescent proteins, producing diffraction-limited spots with an intensity equivalent to 1–3 single fluorescent proteins.

9. For the purpose of this protocol, we are using BSC1 cells stably overexpressing AP2S1-EGFP. AP2S1-EGFP is a well-established marker for clathrin-coated vesicles [10, 17].

10. NaOH dissolves glass, therefore if a glass container is used, after usage remove NaOH 1 M and extensively wash the jar.

11. This protocol promotes the functionalization of glass coverslips. Several variants are possible: NaOH step (Subheading 3.2, **step 7**) might be avoided relaying only on the glass-discharge step (Subheading 3.2 **step 11**) for glass functionalization. Conversely glass functionalization can also obtained by exposing the coverslips to UV light after the washing steps (Subheading 3.2, **step 11**).

12. Since the suggested rack holds up to 8 coverslips the protocol considers the preparation of 8 glass coverslips.

13. The washes are performed by sprinkling the liquid over a coverslip held in forceps at a 30° to 60° angle over an empty beaker.

14. Adding 0.02% Tween 20 improves the spreading of the solution on the glass coverslip.

15. Although contaminations are extremely rare, a wash in 70% ethanol is possible by placing sterile filtered 70% ethanol solution in the wells. The glass coverslips are then inserted in the wells and incubated 10 min in 70% ethanol which is subsequently removed. The coverslips are then rinsed with PBS. Alternatively, the PBS covered coverslips in the six-well plate can be left 30 min under UV illumination.

16. Defining the spreading time is key to obtain the most dynamic cells and the best optical homogenous plasma membrane cell surface. Various fibronectin concentrations should be tested (e.g., 1, 5, 10 nM) to define the concentration that promotes efficient spreading and maintains dynamic clathrin-coated pit formation.

17. The surface of a 25 mm diameter glass coverslip corresponds to $\sim 5 \times 10^7$ μm^2, spreading 2×10^7 beads on the surface corresponds to ~ 0.4 beads/μm^2. In the field of view (~ 2600 μm^2) we might visualize up to ~ 1000 beads. The efficiency of binding may vary within ~ 50–80%.

18. Samples with very high bead concentrations are suboptimal since a majority of beads will tend to cluster or overlap, rendering them unsuitable for the calibration. Moreover, the entire field of view should be sampled in order to capture potential variations in illumination intensity.

19. The acquired images should never have saturated pixels; acquisition parameters must be set accordingly. Since bleaching of fluorescent beads is minimal, we acquire the same field across multiple conditions (e.g., varying exposure time), recording 10–15 consecutive images for each condition. Using this approach, it is possible to directly correlate changes in bead intensity distributions with the tested variable. Diffraction-limited beads should be used (smaller than 200 nm in diameter). If more than one fluorescent channel is used, each laser line should be characterized as described here.

20. It is recommended to analyze these data before continuing with the protocol. The results of the analysis provide the imaging parameters to acquire single molecules and cell dynamics. A "check-up" of the imaging system consisting of this calibration procedure should be performed routinely.

21. Since single fluorophores or fluorescent proteins emit orders of magnitude less fluorescence than beads, only a few scattered beads should be present in the preparation to aid locating the focal plane. Because of competition with proteins for adsorption to the cover glass, we suggest using a concentration corresponding to 20–200 beads per field of view (*see* also **Note 16**) since the binding efficiency drops to 5–10% and only a few beads will effectively adsorb.

22. We suggest detecting the focal plane by focusing on the beads using low laser power to avoid photobleaching the single fluorescent molecules. After detecting the focal plane, slightly move the field of view and acquire a single image to ensure the absence of beads and verify the focus. At this point, the time series can be acquired using the laser power required to visualize single chemical fluorophores or fluorescent proteins.

23. Imaging beads with high laser power can saturate and damage the microchip of EMCCD cameras.

24. The focal plane can be located in TIRF based on single molecules alone. In this case, we recommend avoiding constant illumination to prevent photobleaching. The stage position can be efficiently checked by the acquisition of few consecutive frames, while adjusting the focal plane. Optimal lysate concentration leads to ~1000 fluorescent proteins in the field of view. If the single molecules are too concentrated, their detection can be problematic, leading to a majority of molecules being excluded from further analysis. Conversely, lower concentrations (less than 100molecules/field) will require multiple acquisitions to obtain sufficient statistics.

25. We use an imaging system where the excitation light path passes through the objective. Since the absolute z position of the objective can modify the TIRF angle, we suggest adjusting the focal plane by moving the stage along the z axis and keep the same absolute objective position during the single molecule calibration and the cell imaging experiments.

26. To limit photobleaching, cells should be identified using low laser power; avoid using binocular or continuous TIRF illumination during this step.

27. During clathrin-coated pit-initiation, the first units are recruited every ~2 s [17]. Imaging at 5 Hz or higher frequencies results in ~10 frames between consecutive events, thus enabling the identification of quantal AP2 recruitment to a nascent clathrin-coated pit. At the same illumination power, a higher frequency of acquisition corresponds to a shorter absolute imaging time before photobleaching occurs. In our system, we processed movies with no more than ~20% of bleaching from the first to the last frame of the time-series, which corresponded to ~300 frames. Since a time series of 300 frames at 5 Hz corresponds to 1 min, the acquisition of high-frequency movies should be complemented with low frequency ones (0.5–1 Hz) acquired for 5 min to verify that clathrin-coated vesicles are forming as expected [10, 11, 17].

28. The projection can be limited to the first 5–10 frames to minimize the influence of post-photobleaching frames on the fitting accuracy.

29. When performing automatic detection of single-molecule photobleaching steps, e.g., based on changes of the second derivative [26] or step fitting [17, 27], we recommend inspecting several traces to verify that the automated software is working properly. Noisy traces, traces with more than one step, and traces that belong to molecules separated by less than 400 nm should be discarded.

30. Key parameters for the tracking step are: maximum gap length (2–3 frames, corresponding to "timeWindow" of 3–4); minimum track length (1 frame); minimum/maximum search radius for detections between consecutive frames (3–5 pixels; *see* also Table 1).

31. Although stepwise increases in fluorescence intensity can also be observed past the first 20 s of coated pit assembly, we restricted our analyses to the first two steps because (a) the recruitment of additional AP2-EGFP molecules increases noise between consecutive steps and as a result decreases step detection sensitivity; (b) the growing clathrin lattice pushes the plasma membrane away from the coverslip and into the cytosol, where the TIRF calibration and linearity assumptions are likely invalid.

Acknowledgments

Dr. Cocucci was supported by The Ohio State University Comprehensive Cancer Center (P30-CA016058) and an intramural fund (IRP46050-502339) granted to Dr. Cocucci.

References

1. Kirchhausen T (2000 Dec) Three ways to make a vesicle. Nat Rev Mol Cell Biol 1(3):187–198

2. Kirchhausen T, Owen D, Harrison SC (2014) Molecular structure, function, and dynamics of Clathrin-mediated membrane traffic. Cold Spring Harb Perspect Biol 6:a016725

3. Pearse BMF (1975) Coated vesicles from pig brain: purification and biochemical characterization. J Mol Biol 97(1):93–98

4. Mellman I (1996) Endocytosis and molecular sorting. Annu Rev Cell Dev Biol 12:575–625

5. Saheki Y, De Camilli P (2012) Synaptic vesicle endocytosis. Cold Spring Harb Perspect Biol 4(9):a005645

6. Bittner MA, Aikman RL, Holz RW (2013) A nibbling mechanism for clathrin-mediated retrieval of secretory granule membrane after exocytosis. J Biol Chem 288(13):9177–9188

7. Royle SJ (2012) The cellular functions of clathrin. Cell Mol Life Sci 63(16):1823–1832

8. McMahon HT, Boucrot E (2011) Molecular mechanism and physiological functions of clathrin-mediated endocytosis. Nat Rev Mol Cell Biol 12(8):517–533

9. Gaidarov I, Santini F, R a W, Keen JH (1999) Spatial control of coated-pit dynamics in living cells. Nat Cell Biol 1(1):1–7

10. Ehrlich M, Boll W, Van Oijen A, Hariharan R, Chandran K, Nibert ML et al (2004 Sep 3) Endocytosis by random initiation and stabilization of clathrin-coated pits. Cell 118(5):591–605

11. Aguet F, Antonescu CN, Mettlen M, Schmid SL, Danuser G (2013) Advances in analysis of low signal-to-noise images link dynamin and AP2 to the functions of an endocytic checkpoint. Dev Cell 26(3):279–291

12. Jaqaman K, Loerke D, Mettlen M, Kuwata H, Grinstein S, Schmid SL et al (2008) Robust single-particle tracking in live-cell time-lapse sequences. Nat Methods 5(8)

13. Puthenveedu MA, von Zastrow M (2006) Cargo regulates Clathrin-coated pit dynamics. Cell 127(1):113–124

14. Loerke D, Mettlen M, Schmid SL, Danuser G (2011) Measuring the hierarchy of molecular events during Clathrin-mediated endocytosis. Traffic 12(7):815–825

15. Danuser G (2011) Computer vision in cell biology. Cell 147(5):973–978

16. Cocucci E, Gaudin R, Kirchhausen T (2014) Dynamin recruitment and membrane scission at the neck of a clathrin-coated pit. Mol Biol Cell 25:3595

17. Cocucci E, Aguet F, Boulant S, Kirchhausen T (2012 Aug) The first five seconds in the life of a Clathrin-coated pit. Cell 150(3):495–507

18. Schneckenburger H (2005) Total internal reflection fluorescence microscopy: technical innovations and novel applications. Curr Opin Biotechnol 16(1):13–18

19. Böcking T, Aguet F, Harrison SC, Kirchhausen T (2011 Mar) Single-molecule analysis of a molecular disassemblase reveals the mechanism of Hsc70-driven clathrin uncoating. Nat Struct Mol Biol 18(3):295–301

20. Smith DA (1998) A quantitative method for the detection of edges in noisy time-series. Philos Trans R Soc Lond Ser B Biol Sci 353(June 1997):1969–1981

21. Carter BC, Vershinin M, Gross SP (2008) A comparison of step-detection methods: how well can you do? Biophys J 94(1):306–319

22. Fader CM, Sánchez DG, Mestre MB, Colombo MI (2009) TI-VAMP/VAMP7 and VAMP3/cellubrevin: two v-SNARE proteins involved in specific steps of the autophagy/multivesicular body pathways. Biochim Biophys Acta 1793(12):1901–1916

23. Aggarwal T, Materassi D, Davison R, Hays T, Salapaka M (2012) Detection of steps in single molecule data. Cell Mol Bioeng 5(1):14–31

24. Arunajadai SG, Cheng W (2013) Step detection in single-molecule real time trajectories embedded in correlated noise. PLoS One 8(3):1–9

25. Schwarz G (1978) Estimating the dimension of a model. Ann Stat 6(2):461–464

26. Kerssemakers JWJ, Munteanu EL, Laan L, Noetzel TL, Janson ME, Dogterom M (2006) Assembly dynamics of microtubules at molecular resolution. Nature 442(7103):709–712

27. Calebiro D, Rieken F, Wagner J, Sungkaworn T, Zabel U, Borzi A et al (2013 Jan 8) Single-molecule analysis of fluorescently labeled G-protein-coupled receptors reveals complexes with distinct dynamics and organization. Proc Natl Acad Sci U S A 110(2):743–748

Chapter 16

Cryo-Electron Tomography of the Mammalian Synapse

Rubén Fernández-Busnadiego

Abstract

Characterizing the detailed structure of the mammalian synapse is of crucial importance to understand its mechanisms of function. Here I describe a protocol to study synaptic architecture by cryo-electron tomography (cryo-ET), a powerful electron microscopy technique that enables 3D visualization of unstained, fully hydrated cellular structures at molecular resolution. The protocol focuses on purified synaptic terminals ("synaptosomes"), currently the most suitable preparation to analyze mammalian synaptic architecture by cryo-ET.

Key words Cryo-electron microscopy, Cryo-EM, Neuron, Tissue fractionation, Synaptic vesicle, Postsynaptic density

1 Introduction

Cryo-ET is the technique of choice to study the molecular architecture of the cell at nanometer resolution. To that end the samples need to be vitrified, i.e., frozen at fast cooling rates that prevent the reorganization of water molecules into crystals, and instead form amorphous ice. Vitrified samples are mounted on a cryo-electron microscope capable of keeping them close to liquid nitrogen temperature during imaging. Because no contrasting agents such as heavy metals are applied, the (phase) contrast of these images arises directly from the biological densities. Images of the object of interest are recorded from different angles by tilting the specimen stage, and these images are then computationally recombined into a 3D tomogram. Thus, this technique allows the 3D visualization of fully hydrated, unstained cells in which the fine structural details are preserved to a much greater extent than with classical EM techniques [1].

However, cryo-ET also poses additional challenges in terms of sample preparation. Because the specimens are embedded in vitreous ice, they are much more difficult to section by mechanical means

The original version of this chapter was revised. A correction to this chapter can be found at
https://doi.org/10.1007/978-1-4939-8719-1_19

Laura E. Swan (ed.), *Clathrin-Mediated Endocytosis: Methods and Protocols*, Methods in Molecular Biology, vol. 1847,
https://doi.org/10.1007/978-1-4939-8719-1_16, © Springer Science+Business Media, LLC, part of Springer Nature 2018

than their plastic-embedded counterparts. Vitreous sectioning is nevertheless possible and has been successfully applied to the study of mammalian synapses in organotypic slices [2, 3], but the yield of this technique is low and the vitreous sections suffer from artifacts such as substantial compression along the cutting direction. Dissociated primary neuronal cultures are also amenable for cryo-ET, as they can be grown on EM grids and neurites can be imaged directly without the need of thinning procedures [4, 5]. However, it is hard to find synapses in these preparations as they may preferentially form in areas too thick for direct imaging. It is also possible to study the molecular organization of certain synaptic components using isolated synaptic membranes [6].

Nowadays synaptosomes are arguably the experimental system offering the best compromise between synapse yield and structural preservation for cryo-ET [7]. Synaptosomes consist on portions of the presynaptic and postsynaptic terminals that reseal upon nervous tissue homogenization and are held together by the molecules of the synaptic cleft. These isolated synapses remain largely functional, as they maintain mitochondrial respiration and can be stimulated for Ca^{2+}-dependent neurotransmitter release upon membrane depolarization [8]. However, One should also be aware of the limitations of this system, as synaptosomes cannot carry out sustained neurotransmitter release and delicate structural features such as the actin cytoskeleton may be altered during preparation. Nevertheless, our studies showed that synaptic architecture in synaptosomes was not only structurally comparable to that of tissue slices [2], but also a proxy for the electrophysiological properties of the brain [9]. In the following I present a detailed protocol on the extraction and vitrification of murine synaptosomes for cryo-ET studies.

2 Materials

2.1 Synaptosome Extraction

For all buffers use ultrapure water and analytical grade reagents. The amounts given here are for one mouse brain, which provides many times more material than needed in a typical cryo-ET experiment. The protocol is based on the Percoll gradient method [10–12].

1. Sucrose buffer (5×): 5 mM EDTA, 1.25 mM DL-DTT, 20 mM HEPES, 1.6 M sucrose. Prepare the following solutions separately: 0.25 M EDTA (dissolve 2.33 g in 25 ml of water; add NaOH pellets to allow dissolution and adjust pH to 7.4) and 0.1 M DL-DTT (dissolve 0.15 g in 10 ml of water). Dissolve 2.38 g of HEPES in 400 ml, adjust pH to 7.4, and then add 273.84 g of sucrose (stir for ~1 h at 30 °C to allow dissolution). Add 10 ml of 0.25 M EDTA and 6.25 ml of 0.1 M DL-DTT and adjust the volume to 500 ml with water. This solution can be stored at −80 °C. Make 50 ml aliquots and freeze.

2. Homogenization buffer: Dissolve 5 cOmplete mini EDTA-free protease inhibitor tablets (Roche) in 7 ml of water. Immediately before use add 63 ml of 1× sucrose buffer.

3. Percoll solutions: For 3%/10%/23% Percoll solutions mix 10 ml 5× sucrose buffer respectively with 1.5/5/11.5 ml Percoll and adjust the volume to 40 ml with water. Adjust pH to 7.4 at 4 °C and bring to 50 ml with water. Work with Percoll in sterile conditions. These solutions can be stored for a week at 4 °C.

4. Percoll gradients: Lay 2 ml of 3% Percoll solution with a pipette. Wash peristaltic pump with 10% Percoll solution and load 2 ml into the tube by pressing a glass pipette coupled to the pump against the bottom of the tube. Repeat this step for the 23% Percoll solution. Make two gradients.

5. Saline buffer: 140 mM NaCl, 5 mM KCl, 5 mM NaHCO$_3$, 1.2 mM NaH$_2$PO$_4$, 1 mM MgCl$_2$, 10 mM glucose, 10 mM HEPES. For 500 ml, add the following amounts to 400 ml of water: 4.1 g NaCl, 0.19 g KCl, 0.21 g NaHCO$_3$, 0.08 g NaH$_2$PO$_4$·H$_2$O, 0.10 g MgCl$_2$·6H$_2$O, 0.9 g glucose, 1.19 g HEPES. Adjust pH to 7.4 and bring to 500 ml with water. This solution can be stored for a week at 4 °C.

6. Murine dissection surgical kit.

7. Teflon-glass homogenizer.

2.2 Synaptosome Vitrification

1. Copper EM grids coated with a holey carbon support.

2. 10 nm BSA-coated gold fiducial markers (Aurion).

3. Whatman #1 filter paper.

---

3 Methods

3.1 Synaptosome Extraction

It is critical to perform this procedure at 4 °C. Always work on ice using precooled tools, tubes and solutions.

1. Remove the brain from the skull and rinse in homogenization buffer.

2. Remove the cerebellum. Separate the brain hemispheres and carefully remove the white matter and blood vessels.

3. Homogenize in 4 ml of homogenization buffer (*see* **Notes 1** and **2**).

4. Load the homogenate in a 50 ml tube and centrifuge for 2 min at 3000 × *g* (5000 rpm on an SS-34 rotor). Pour the supernatant (S1) into a new centrifuge tube making sure that no pellet (P1) is transferred.

5. Gently rock the tube alternatively clockwise and counter-clockwise to detach P1 from the tube wall and resuspend care-

fully using a plastic pipette (*see* **Note 3**). Balance with homogenization buffer and centrifuge again for 2 min at 3000 × *g* (5000 rpm on an SS-34 rotor).

6. Pour supernatant (S1') into a new centrifuge tube making sure that no pellet (P1') is transferred. Combine S1 and S1' and centrifuge for 15 min at 9200 × *g* (8800 rpm on an SS-34 rotor).

7. Remove the supernatant (S2) by aspiration and carefully resuspend the pellet (P2) in 3 ml sucrose buffer.

8. Load the Percoll gradients with resuspended P2 very slowly to avoid disturbing the gradients. Centrifuge 12 min at 18700 × *g* (12,500 rpm on an SS-34 rotor).

9. Slowly remove the synaptosome band at the 10–23% Percoll interface using a glass pipette and place in a new centrifuge tube.

10. Fill up each tube with 30 ml saline solution and centrifuge for 12 min at 18700 × *g* (12,500 rpm on an SS-34 rotor) to wash the Percoll away.

11. Remove the supernatant (S3) very carefully and resuspend the pellet (P3) in as little saline solution as possible.

12. Determine protein concentration and dilute in saline solution as needed to obtain 1 mg/ml. Keep at 4 °C.

3.2 Synaptosome Vitrification by Plunge Freezing

Given their relatively small size (typically a few hundred nanometers in diameter), synaptosomes can be successfully vitrified by plunge freezing into a fast cooling cryogen such as liquid ethane cooled down to liquid nitrogen temperature.

1. It is beneficial to mix gold fiducial particles with synaptosomes to aid the computational alignment of tomographic projections [13]. BSA-coated gold fiducials minimize aggregation and interactions with the sample. The fiducials need to be centrifuged prior to mixing with synaptosomes to (1) remove the sodium azide used as preservative, and (2) achieve the desired final concentration. Spin 125 µl of 10 nm BSA-coated gold fiducial markers for 90 min at 15,500 × *g* using a tabletop centrifuge. Carefully remove ~100 µl of supernatant without disturbing the pellet so that the remaining volume is ~ 25 µl (5× concentrated gold particles). Resuspend with another 25 µl of saline buffer and incubate at 37 °C.

2. Glow discharge the EM grids using a plasma cleaner to render them hydrophilic.

3. Incubate synaptosomes at 37 °C in a water bath for 30 min prior to vitrification so that they regain their physiological properties. During this incubation it is also possible to apply pharmacological treatments of interest.

4. Mix 50 μl of gold fiducials resuspended in saline buffer with 50 μl of synaptosomes.

5. Mount a glow-discharged EM grid in the vitrification device and apply 3 μl of the synaptosome–gold fiducial mixture on the side of the grid coated by the carbon film. Incubate 10 s to allow synaptosomes to settle on the carbon film of the EM grid.

6. Blot the excess buffer from the grid using Whatman #1 filter paper. If the vitrification device allows single-sided blotting, blot only from the back of the grid (the opposite side where the sample was deposited). Plunge immediately into the cryogen, e.g., liquid ethane or 1:2 ethane–propane mixture (*see* **Note 4**).

3.3 Cryo-ET Imaging of Synaptosomes

Synaptosomes can be imaged by cryo-ET using any cryo-electron microscope equipped with a tilting stage and automated tomography software such as SerialEM [14]. For best results, use a 300 kV microscope equipped with a field emission gun, energy filter and a direct electron detector.

1. Load the EM grid containing vitrified synaptosomes in the cryo-electron microscope. Using low magnification (~ 125×) identify regions of suitable ice thickness (*see* **Note 5**).

2. The most abundant feature in this preparation is isolated presynaptic terminals without an associated postsynaptic counterpart (Fig. 1a, b, black arrowheads), which may not be suitable for studies on synaptic transmission (*see* **Note 6**). Cellular debris and isolated mitochondria can also be seen. Fig. 1c and d show a typical synaptosome consisting on a presynaptic terminal associated to a much smaller postsynaptic compartment. Presynaptic terminals in synaptosomes often contain hundreds of synaptic vesicles (some of which may be clathrin-coated (Fig. 1e)), mitochondria, and ER membranes.

3. For optimal tomogram quality, synaptosomes should be embedded in a layer of ice as thin as possible, surrounded by sufficient fiducial markers (*see* **Note 7**), placed within the holes of the carbon film and in areas allowing a tilt range of at least ±60 °.

4   Notes

1. Homogenization must be careful to minimize heating. For the first stroke, the homogenizer should be started at minimum speed once the piston is inside the solution. After this stroke the speed should be increased slowly to 900 rpm and homogenization should proceed until no small parts of tissue are visible (seven strokes maximum). When homogenization is

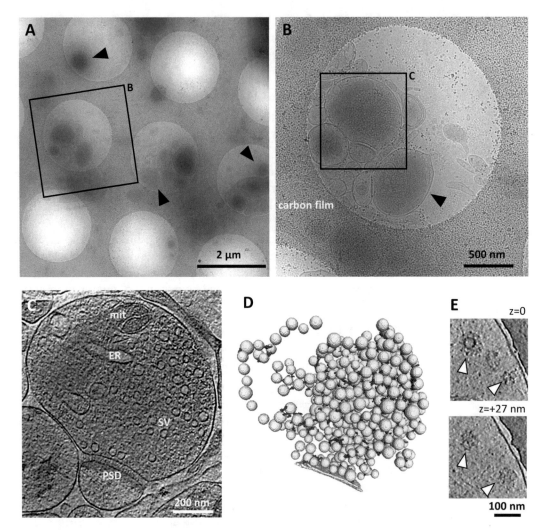

Fig. 1 Cryo-ET imaging of rat synaptosomes. (**a, b**) 2D images of increasing magnification of an EM grid coated with a holey carbon film containing vitrified frozen-hydrated rat synaptosomes. Black arrowheads mark isolated presynaptic terminals without a postsynaptic counterpart. (**c**) Tomographic slice of the synaptosome marked in B. The presynaptic terminal contains synaptic vesicles (SV), a mitochondrion (mit), and ER membranes (ER). On the postsynaptic terminal the postsynaptic density (PSD) is visible. (**d**) Semiautomated 3D rendering of the area within 250 nm of the active zone of the synaptosome in C, showing synaptic vesicles (yellow), the active zone (grey), and the fine filaments linking synaptic vesicles to each other (red) and to the active zone (blue). (**e**) Tomographic slices of another synaptosome depicting the same region in x and y at different z heights (z = 0 nm and z = + 27 nm respectively). White arrowheads mark a clathrin-coated synaptic vesicle (top) and an empty clathrin coat (bottom). Tomographic slices are 2.7 nm-thick (modified from Ref. [2] with permission of Rockefeller University Press)

completed, reduce the speed to minimum and remove the piston from the solution.

2. Rapid dissection and homogenization are crucial for the purification of functional synaptosomes. If more than one brain is being

used, the procedure can be paused at this point and repeated for subsequent brains. The homogenates can then be pooled.

3. Gentle resuspension of the pellets is mandatory to preserve synaptosome integrity. The pellet can be detached from the tube by softly rocking it alternatively clockwise and counterclockwise. Take up the pellet with a pipette and flush it out, making sure that the pipette tip makes tight contact with the bottom of the tube to maximize the efficiency of the resuspension and minimize the number of iterations (The process should be repeated until no visible pieces are visible, ideally not more than three times.). If the pellet to be resuspended is larger than one pipette volume, use fresh tubes for resuspension to avoid disturbing already resuspended material.

4. Whereas pure ethane solidifies at liquid nitrogen temperature; a 1:2 ethane-propane mixture remains liquid and still allows for good vitrification [15]. However, this mixture can also be liquid above the vitrification temperature, so it must be cooled continuously by liquid nitrogen to ensure proper freezing.

5. Synaptosomes tend to form large clumps that are too thick for direct cryo-ET imaging. Appropriate resuspension prior to freezing can alleviate this problem.

6. To speed up the search of synaptosomes suitable for cryo-ET imaging, correlative cryo-light microscopy [16–18] can be applied provided that synaptosomes have been loaded with fluorescent dyes (e.g., FM1-43) or were extracted from animals expressing fluorescently labeled synaptic proteins.

7. Fiducial gold markers allow for optimal alignment of tomographic projections, but current feature-based alignment methods such as those implemented in IMOD [13] also provide good results.

Acknowledgments

I wish to thank Eri Sakata for the critical reading of the manuscript. R. F.-B. is supported by the FP7 GA ERC-2012-SyG_318987–ToPAG grant from the European Commission.

References

1. Asano S, Engel BD, Baumeister W (2015a) In situ Cryo-Electron tomography: a postreductionist approach to structural biology. J Mol Biol

2. Fernandez-Busnadiego R, Zuber B, Maurer UE, Cyrklaff M, Baumeister W, Lucic V (2010) Quantitative analysis of the native presynaptic cytomatrix by cryoelectron tomography. J Cell Biol 188:145–156

3. Zuber B, Nikonenko I, Klauser P, Muller D, Dubochet J (2005) The mammalian central nervous synaptic cleft contains a high density of periodically organized complexes. Proc Natl Acad Sci U S A 102:19192–19197

4. Asano S, Fukuda Y, Beck F, Aufderheide A, Forster F, Danev R, Baumeister W (2015b) A molecular census of 26S proteasomes in intact neurons. Science 347:439–442

5. Lucic V, Kossel AH, Yang T, Bonhoeffer T, Baumeister W, Sartori A (2007) Multiscale imaging of neurons grown in culture: from light microscopy to cryo-electron tomography. J Struct Biol 160:146–156

6. Zuber B, Unwin N (2013) Structure and superorganization of acetylcholine receptor-rapsyn complexes. Proc Natl Acad Sci U S A 110:10622–10627

7. Fernandez-Busnadiego R, Schrod N, Kochovski Z, Asano S, Vanhecke D, Baumeister W, Lucic V (2011) Insights into the molecular organization of the neuron by cryo-electron tomography. J Electron Microsc 60(Suppl 1):S137–S148

8. Whittaker VP (1993) Thirty years of synaptosome research. J Neurocytol 22:735–742

9. Fernandez-Busnadiego R, Asano S, Oprisoreanu AM, Sakata E, Doengi M, Kochovski Z, Zurner M, Stein V, Schoch S, Baumeister W et al (2013) Cryo-electron tomography reveals a critical role of RIM1alpha in synaptic vesicle tethering. J Cell Biol 201:725–740

10. Dunkley PR, Heath JW, Harrison SM, Jarvie PE, Glenfield PJ, Rostas JA (1988) A rapid Percoll gradient procedure for isolation of synaptosomes directly from an S1 fraction: homogeneity and morphology of subcellular fractions. Brain Res 441:59–71

11. Dunkley PR, Jarvie PE, Robinson PJ (2008) A rapid Percoll gradient procedure for preparation of synaptosomes. Nat Protoc 3:1718–1728

12. Godino MD, Torres M, Sanchez-Prieto J (2007) CB1 receptors diminish both Ca^{2+} influx and glutamate release through two different mechanisms active in distinct populations of cerebrocortical nerve terminals. J Neurochem 101:1471–1482

13. Kremer JR, Mastronarde DN, McIntosh JR (1996) Computer visualization of three-dimensional image data using IMOD. J Struct Biol 116:71–76

14. Mastronarde DN (2005) Automated electron microscope tomography using robust prediction of specimen movements. J Struct Biol 152:36–51

15. Tivol WF, Briegel A, Jensen GJ (2008) An improved cry ogen for plunge freezing. Microsc Microanal 14:375–379

16. Fukuda Y, Schrod N, Schaffer M, Feng LR, Baumeister W, Lucic V (2014) Coordinate transformation based cryo-correlative methods for electron tomography and focused ion beam milling. Ultramicroscopy 143:15–23

17. Schellenberger P, Kaufmann R, Siebert CA, Hagen C, Wodrich H, Grunewald K (2014) High-precision correlative fluorescence and electron cryo microscopy using two independent alignment markers. Ultramicroscopy 143:41–51

18. Zhang P (2013) Correlative cryo-electron tomography and optical microscopy of cells. Curr Opin Struct Biol 23:763–770

Chapter 17

Quantitative Analysis of Clathrin-Mediated Endocytosis in Yeast by Live Cell Fluorescence Microscopy

Eric B. Lewellyn and Yansong Miao

Abstract

The budding yeast *Saccharomyces cerevisiae* has provided a useful model for studying clathrin-mediated endocytosis due to ease of genetic manipulation and crosssectional imaging of individual endocytic sites. This protocol describes a method for using live cell fluorescence microscopy to analyze clathrin-mediated endocytosis and the contributions of actin to the process.

Key words *Saccharomyces cerevisiae*, Clathrin, Endocytosis, Actin, Myosin, Fluorescence microscopy, *SLA1*, *ABP1*

1 Introduction

Clathrin-mediated endocytosis (CME) is an essential pathway for trafficking plasma membrane lipids, proteins, and external macromolecules into the endomembrane system (for review [1]). The budding yeast *Saccharomyces cerevisiae* has proven to be a highly valuable model to study CME because the pathway is highly conserved among eukaryotes (for reviews, *see* [2, 3]). Studies in *S. cerevisiae* have revealed that endocytic proteins are grouped into functional modules with distinct spatial and temporal recruitment [4], which are also apparent in mammalian cells [2]. Importantly, actin and myosins play an essential role in driving membrane invagination in yeast [5–7] as well as those mammalian cells with membranes under tension [8].

These discoveries have been facilitated by straightforward genetic manipulation in *S. cerevisiae* coupled with fluorescence microscopy-based assays for CME. Tagging endogenous copies of endocytic proteins with enhanced green fluorescent protein (GFP) or other fluorescent protein tags can be achieved by homology-directed gene editing. Molecular tool kits have been developed that simplify the

The original version of this chapter was revised. A correction to this chapter can be found at
https://doi.org/10.1007/978-1-4939-8719-1_19

Laura E. Swan (ed.), *Clathrin-Mediated Endocytosis: Methods and Protocols*, Methods in Molecular Biology, vol. 1847,
https://doi.org/10.1007/978-1-4939-8719-1_17, © Springer Science+Business Media, LLC, part of Springer Nature 2018

process by providing plasmid templates with fluorescent tags and selection cassettes [9–11]. Fluorescently tagging endocytic proteins using these techniques has revealed that individual endocytic proteins are recruited to diffraction-limited micro domains called endocytic patches with regular timing [4]. Mutations that affect the endocytic pathway change the intensity, duration, and movement of labeled endocytic patch proteins and provide a quantitative assay for interpreting how mutations affect the process [4, 12].

This protocol describes a basic epifluorescence microscopy assay for endocytosis and data processing using ImageJ/FIJI [13]. Budding yeast cells are roughly spherical so imaging in a medial focal plane provides cross-sectional view of endocytic buds. A translational fusion between coat protein Sla1 and green fluorescent protein (Sla1-GFP) provides a marker for the endocytic coat while the actin-binding protein Abp1, when translationally fused to monomeric red fluorescent protein (Abp1-mRFP), is a bright marker for the actin network [7, 14]. These proteins are recruited to diffraction-limited patches at the cortex. If an individual patch is imaged over time, the temporal differences in recruitment of each endocytic protein can be observed through two-color fluorescence microscopy [12]. Mutations in many genes in the endocytic pathway alter the lifetimes of individual patch proteins, which can be quantitatively analyzed through live cell fluorescence microscopy.

2   Materials

All reagents should be prepared using highly pure water.

1. Imaging media: For 500 mL, add about 300 mL water to the beaker with stirring. Weigh 10 g D-glucose, 3.35 g yeast nitrogen base without amino acids. If the yeast are auxotrophic, any required supplements must be added to the media to support growth. For most amino acids and nucleobase precursors, add 50 mg of each per 500 mL solution. If required, adenine sulfate supplementation requires 10 mg per 500 mL solution, and L-leucine requires 90 mg per 500 mL solution (*see* **Note 1**). Dissolved reagents can be sterilized by filtration through a filter with 0.22 μm or smaller pore size.

2. 1 mg/mL Concanavalin A (ConA) solution: To produce a 25 mL solution, measure 25 mL of phosphate-buffered saline supplemented with $CaCl_2$ and $MnCl_2$ (137 mM NaCl, 2.7 mM KCl, 10 mM Na_2HPO_4, 2.5 mM $CaCl_2$, 2.5 mM $MnCl_2$. Adjust the pH to 6.8 with 1 N HCl). Dissolve 25 mg ConA type IV powder from *Canavalia ensiformis* (jack bean). Filter sterilization is not necessary if cells are going to be imaged immediately. Generate 0.2 mL Aliquots and store at −20 °C. This solution can be thawed and refrozen at least five times without any noticeable decrease in cell adsorption.

3. YPD Plates: 2% (W/V) Bacto peptone, 1% (W/V) Bacto yeast extract, 2% (W/V) D-glucose, 2% (W/V) agar. For 1 L, dissolve 20 g Bacto peptone, 10 g Bacto yeast extract, and 20 g Bacto agar. Dissolve reagents in 0.7 L pure water with stirring. Bring total volume to 0.9 L and sterilize by autoclaving. Separately, dissolve 20 g D-glucose in 100 mL water and sterilize by filtration through a filter with 0.22 μm or smaller pore size. After autoclaving, mix both solutions and pour into 10 cm plastic petri dishes. Pour 20 mL per dish. Store the plates at 4° C or room temperature.

4. Cell line stably expressing a construct of interest: This protocol uses cell expressing Sla1-GFP and Abp1-mRFP, strain DDY3982 [5].

5. Slides and coverslip: Standard prewashed 75 mm × 25 mm × 1 mm glass slides of any make can be used. For coverslips, careful attention should be paid to the thickness of the coverslip. Most high NA objectives are designed to be matched with coverslip no. 1.5H, which signifies 0.170 +/− 0.005 mm thick borosilicate glass.

6. Immersion fluid: Low autofluorescence immersion oil should be used for immersion lenses and the refractive index should be matched to the experimental conditions (see **Note 2**).

7. Vacuum grease pen: This can be generated using a 10 cc syringe with Luer taper or Luer lock end filled with silicon vacuum grease (Dow Corning Corporation, Midland, MI, USA or equivalent) (see **Note 3**).

8. Microscope: Either epifluorescence or scanning confocal microscope (see **Note 4**) 60×–100× PLANO objectives with 1.4NA. Light source (LED light is preferable).

9. Parafilm.

10. Temperature-controlled shaking incubator at 25C (This needs to be made dark.).

11. Spectrophotometer.

3 Methods

3.1 Slide Preparation

For all imaging procedures, protect the yeast from excitation light as much as possible to minimize fluorescence bleaching.

1. Maintain cells on YPD plates. Cells from freezer stocks should also be revived on YPD plates prior to imaging (see **Note 5**). Most yeast strains grow fastest at 30° C, however, at 25° C, cells are less likely to accumulate compensatory mutations if there are any defects in the endocytic pathway. It is therefore advisable that all strains be routinely maintained at 25° C.

2. Inoculate a starter culture consisting of 3 mL of imaging media and the yeast strain to be imaged 1 day before imaging. The starter culture should be inoculated to an optical density of about 0.05 absorbance at 600 nm (O.D.$_{600}$) (*see* **Note 6**). Incubate in a dark shaking incubator at 25 °C overnight. On the day of imaging, determine the O.D.$_{600}$ of the overnight culture and inoculate 3 mL of fresh imaging media to an O.D.$_{600}$ of 0.1 or 0.2 if slow-growing strains are to be imaged. Cells should be imaged in mid log-phase growth, when the OD$_{600}$ of the culture is between 0.4 and 0.6; this takes about 5 h for healthy strains.

3. Coat the coverslip with ConA so the cells will adsorb to the surface. Place the coverslip no. 1.5H on a sheet of Parafilm and pipet 200 μL of 1 mg/mL ConA solution to the center (Fig. 1a). Allow the ConA to adsorb to the glass surface for 10 min (*see* **Note 7**). Aspirate the ConA solution from the coverslip and wash twice with imaging media by pipetting 400 μL imaging media onto the coverslip and removing by aspiration (*see* **Note 8**).

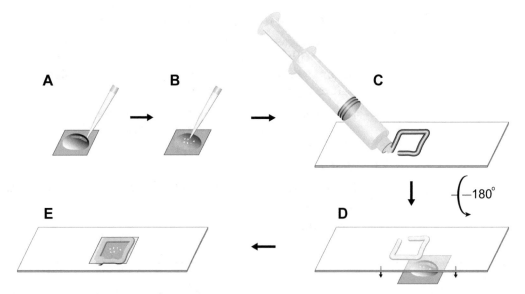

Final corner sealed to complete slide   Slide inverted and pressed onto coverglass

Fig. 1 Assembly of a live cell imaging slide. (**a**) Place a coverslip with thickness 0.17 mm +/− 0.005 mm on a sheet of Parafilm and pipet 200 μL of 1 mg/mL ConA solution to the top to adsorb for 10 min. Then wash the coverslip with 400 μL imaging media and aspirate. (**b**) Add media containing suspended cells to the slide to adsorb for 10 min. Wash away unattached cells and cover the slide with 50 μL of fresh imaging media. (**c**) Place a bead of vacuum grease on a slide to create the chamber for imaging media. Leave one corner is open for the air to escape. The sides of the bead should be approximately 2 mm from the edge of the cover glass. (**d**) Invert the slide on the cover glass and press out and excess air and imaging media. (**e**) Press a small amount of additional vacuum grease in from the side of the chamber to seal the remaining corner of the slide

4. Pipet 400 μL of suspended cells (OD_{600} 0.4–0.6) onto the coverslip and incubate at room temperature for 10 min.

5. Aspirate the media and wash the cells that have not attached by gently pipetting 400 μL of imaging media up and down across the attached cells about 5–6 times. Then, aspirate the media with the dislodged cells (*see* **Note 9**). Repeat this twice to ensure that only firmly attached cells remain on the coverslip. Finally, add 50 μL of fresh imaging media to the coverslip to cover the cells (Fig. 1b).

6. Use a vacuum grease pen to make an imaging chamber. Apply a square bead of vacuum grease to the microscope slide such that center of the bead is about 2 mm inside the outer edge of the coverslip (Fig. 1c). Be sure to leave a gap in the bead for excess air and media to escape.

7. Invert the slide and gently press onto the coverslip with adsorbed cells and 50 μL imaging media. Air will escape through the open corner in the vacuum grease bead until the liquid contacts the slide.

8. Once the excess air has been pressed out of the chamber, use a pipet tip to press a small amount of extra slide grease into the open corner to seal the opening (*see* **Note 10**).

3.2 Image Acquisition and Analysis

1. Input acquisition parameters including frame rate and light intensity into the microscope and camera control software. For this protocol, we imaged cells expressing Sla1-GFP and Abp1-mRFP using a Leica TCS SP5 II laser scanning confocal microscope. This microscope uses an acousto-optical beam splitter and two Leica HyD detectors for detection, and we found a frame rate of 1.0 frame per second was ideal. However, different yeast strains and imaging equipment may require different image acquisition parameters (*see* **Note 11**). The intensity of epifluorescence illumination light should be set such that images can be clearly visualized throughout the movie, but photobleaching is minimized. This information should be established in a pilot experiment before data collection for the actual experiment.

2. Using low power transmitted light illumination, set the focus on a medial focal plane of the cells so endocytic sites will be imaged in cross section (Fig. 2a) (*see* **Note 12**). Select a spot where cells are at least several μm from neighboring cells and begin fluorescence image capturing (*see* **Note 13**).

3. Convert images into a hyperstack in ImageJ/FIJI software. Assign a false color (*see* **Note 14**) to the monochrome image files corresponding to each color channel (Fig. 2b). The ImageJ/FIJI user guide provides specific instructions on basic software operation including how to combine TIFF images into a colorized hyperstack. (For details, *see* https://imagej.nih.gov/ij/

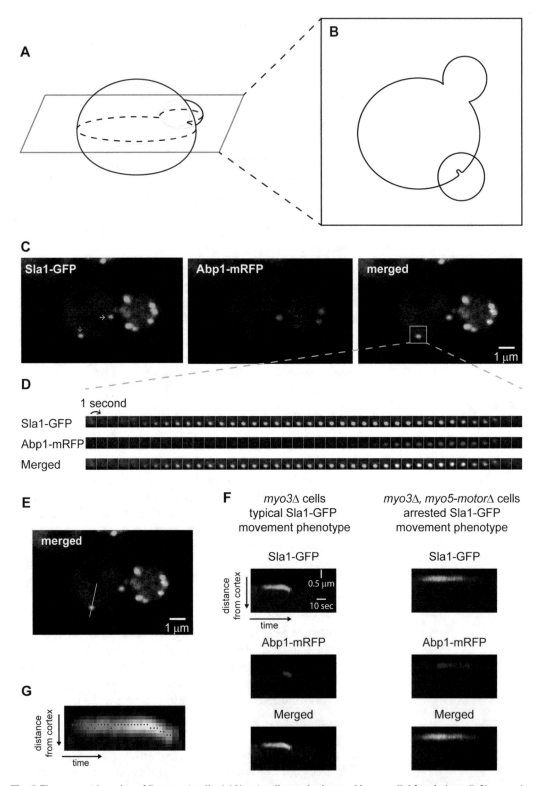

Fig. 2 Fluorescent imaging of live yeast cells. (**a**) Yeast cells can be imaged in a medial focal plane (left), revealing endocytic buds in cross section (right, circle). (**b**) Fluorescent images of yeast cells with coat module protein Sla1 labeled with GFP (green), actin module protein Abp1 labeled with mRFP (magenta), and a merged

docs/guide/user-guide.pdf). In the example shown, Sla1-GFP and Abp1-mRFP appear as diffraction limited spots on the cell cortex (Fig. 2c). Sla1-GFP is recruited first followed by Abp1-mRFP. Then both proteins disappear (Fig. 2d).

4. To analyze individual patches, we recommend generating a kymograph in ImageJ/FIJI. Draw a radial line originating at the center of the cell that passes through the middle of the endocytic patch to be analyzed (Fig. 2e). Generate a kymograph using the ImageJ/FIJI reslice function to create the kymograph (https://imagej.nih.gov/ij/docs/guide/user-guide.pdf) (*see* **Note 15**). The resulting kymograph renders the endocytic patch as a two dimensional representation in which the *x*-axis is time and the *y*-axis is the pixel intensity along the length of the line (Fig. 2f).

5. To assess the recruitment kinetics, measure the lifetime of Sla1-GFP, Abp1-mRFP, and the time difference between the first appearance of each by viewing as a montage (Fig. 2d) or a kymograph (Fig. 2f). Note the difference between the recruitment kinetic for a cell with wild-type endocytosis and the cell with *myo3Δ* and *myo5-motorΔ* mutant alleles (Fig. 2f). Because endocytic proteins have very regular recruitment timing, a similar approach can be used to analyze most combinations of endocytic proteins.

6. To determine whether an endocytic patch is invaginating and undergoing scission, follow the path of the brightest pixel in the kymograph (Fig. 2g). If the path moves away from the cortex without retracting, the endocytic patch is believed to have completed invagination and scission [12, 15]. Endocytic mutants, such as the *myo3Δ myo5-motorΔ* mutant, greatly reduce the frequency with which patches undergo invagination and scission (Fig. 2f) [5] (*see* **Note 16**).

Fig. 2 (continued) image at right. The mother cell contains two representative endocytic patches as indicated by arrows (↓ and →). The lower patch, ↓, is at an earlier stage of endocytosis and Abp1 recruitment has not yet occurred. The patch to the right, → is at a later stage of endocytosis and Abp1 is present, indicating actin polymerization has begun. (**c**) As yeast cells are imaged over time, endocytic patch proteins appear and disappear with characteristic timing. (**d**) A time lapse montage of the boxed area shows the differences between Sla1-GFP and Abp1-mRFP at an endocytic patch over time, with Sla1-GFP recruitment occurring first and Abp1-mRFP recruitment occurring as the patch movement begins. (**e**) ImageJ/FIJI can generate a kymograph using the resliced function. In this example, a radial line through the middle of an endocytic patch was used (line) to generate kymographs. (**f**) Representative kymographs from the cell in part E (left) and an endocytic mutant cell lacking functional *MYO3* and *MYO5* genes. All images were generated for this chapter using a Leica TCS SP5 II laser scanning confocal microscope with excitation by 488 nm and 594 nm lasers and HyD detectors for simultaneous imaging of both channels. Each pixel is 58.8 nm and frames were collected every 0.5 s and binned to 1 frame per second. Strains imaged were DDY3982 and DDY4783 [5]. (**g**) A close up of the kymograph to illustrate the path taken by the brightest pixel. The brightest pixel at each time point is highlighted in red. The patch moves about three pixels inward from the cortex before disappearing. Each pixel is 58.8 nm so the endocytic patch shown internalized ~176 nm

4 Notes

1. Most experimental strains are auxotrophic and require essential nutrient supplementation. However, L-tryptophan should be avoided if possible as it contributes to background fluorescence when GFP is imaged. As an alternative to supplementing with individual nutrients, the imaging media can be supplemented with 2% (W/V) casamino acid powder. The casamino acid supplement can enhance growth for strains with extremely low viability, but may contribute to slightly higher background fluorescence.

2. The light emitted from a fluorescent sample will produce a characteristic diffuse pattern of light known as a point spread function (PSF). When viewed in the z-dimension, the PSF forms an hourglass pattern that is symmetric above and below the plane of focus [16]. An asymmetric PSF is known as circular aberration and it will occur if there is a mismatch between the objective, coverslip thickness, focal depth, and immersion fluid index of refraction. Because focal depth and imaging temperature are subject to experimental requirements, the immersion fluid should be empirically optimized for the experimental conditions. Immersion fluids of various refractive indices ranging from 1.500 to 1.534 are available from Cargille-Sacher Laboratories (Cedar Grove, NJ) or GE Healthcare (Chicago, IL). A refractive index of 1.518 at 23 °C is a good starting point, and the oil with the lowest possible auto fluorescence should be selected. For a more detailed description of circular aberration and immersion oil optimization, *see* [16].

3. To prepare a grease pen, remove plunger from the back of the syringe and firmly press the opening of the vacuum grease tube against the back of the syringe. Squeeze vacuum grease into the syringe until the grease is 0.5 cm from the tip of the syringe. Replace the plunger and press all remaining air out of the front of the syringe. Take care to avoid bubbles in the vacuum grease if possible.

4. Considerations for selecting a microscope:

   (a) *Microscopy setup.* Imaging endocytic patches requires an imaging setup that is optimized for resolution and light capture. First, an oil immersion objective with 60×–100× magnification is necessary to resolve nearby patches and observe the movement of the centroid or local maxima of each patch. Objective lenses should also have flat field correction to prevent distortion of the imaging field, as indicated by "Plan" on the side of the objective for most manufacturers. Objectives should also not employ phase contrast or other optical modifications that decrease the amount of transmitted fluorescent light. Finally, a high numerical aperture

(NA) is necessary to maximize the light captured and increase the precision of endocytic patch localization. Ideally the NA should be 1.4 or greater, although objectives with a lower NA may also provide adequate light capture for quantitative analysis. If two or more colors are to be imaged, the objective should also have the highest level of correction for chromatic aberration. Chromatic aberration is the tendency of lower wavelength light to diffract more than higher wavelength light. For most microscope manufacturers, the highest available correction is indicated by "Apo" on the side of the objective. These objectives typically reduce chromatic aberration to below the limit of resolution across the viewing field, although additional post processing correction may be required if a higher level of alignment precision is required or if elements in the light path introduce additional chromatic aberration.

(b) *Selection of Laser Scanning Confocal or Standard Epifluorescence microscopy.* Endocytic patches can be imaged using either a standard epifluorescence microscope or a fast laser scanning confocal microscope. Confocal imaging increases the axial resolution (resolution in the z-dimension), so endocytic patches above and below the focal plane will be observed less than with standard epifluorescence microscopy. These out-of-plane patches may contribute to a systematic underrepresentation of the distance a local maximum or centroid migrates, although the magnitude of error caused by out-of-plane patches is estimated to be no more than 10% [17]. Also, if a standard epi-fluorescence microscope is used, the same systematic underrepresentation will be present in both the experimental and control samples so the overall conclusions of the assay are not likely to change. A final consideration is that many confocal microscopes are not able to capture images across a large field of view at the same frame rate as a standard epifluorescence microscope. Therefore, the benefits of confocal microscopy should be carefully weighted against these drawbacks.

(c) *Image Digitization.* Analysis of imaging data requires that digital micrographs be obtained with sufficient sensitivity to detect fluorescent endocytic markers and sufficient resolution to discriminate between nearby, diffraction-limited endocytic patches. In a standard epifluorescence microscopy setup, high sensitivity detection typically requires a purpose-built microscope camera with a high quantum efficiency and very low background signal. Examples of cameras that meet these requirements include the Flash 4.0v3 or Orca-R2C 10600-10B by Hamamatsu (Hamamatsu City, Japan) that employs electron multiplying charge coupled device (EMCCD) detector and the Neo 5.5 by Andor (Belfast, UK) that utilizes a scientific complementary metal–oxide–semiconductor (SCMOS) detector. The high sensitivity of these and similar cameras comes at the expense of color detection, so color discrimination must be achieved through filters in the light path. To prevent loss of positional information during image capture, the camera must also have a sufficient pixel density to faithfully reproduce the analog image. According to the Nyquist sampling criterion for lossless digitization, the sampling frequency should be two times the smallest frequency to be resolved [18]. In microscopy, the smallest frequency that can be resolved is the Abbe/Rayleigh limit of resolution (d_{min}), which is determined by both the wavelength of emitted light (λ) and the objective NA:

$$d_{min} = 0.61\lambda/\text{NA}$$

Because a two-dimensional image is being digitized using a square grid of pixels, the sampling frequency should be further increased to 2.4 times the smallest frequency to account for the fact that some points will be diagonally oriented relative to the x-y axes of the pixel array [18]. If laser scanning confocal microscopy is used for image capture, a highly sensitive detector, such as the Leica hybrid detector HyD (Wetzlar, Germany) or a gallium arsenide phosphide (GaAsP) detector is necessary for sensitive signal detection and a high signal-to-noise ratio. The detector measures the total light emitted from the sample as the excitation laser scans across the sample and extrapolates the point sources of fluorescence based on the position of the excitation laser. The resulting image consists of virtual pixels, with the size and number of pixels specified by the software. Because the pixel size is not fixed, lower magnification objectives, like a 63×, 1.4 NA objective can be used without sacrificing resolution. A 63× objective may even be preferable to a high magnification objective because it allows for collection of more photons. For a more detailed description of localization precision, *see* http://zeiss-campus.magnet.fsu.edu/articles/superresolution/palm/practicalaspects.html. Some over sampling is desirable to insure faithful digitization of the signal, but excessive over sampling reduces brightness and increases file size. Ultimately, the pixel size should be experimentally optimized to produce the most precise positional resolution.

(d) *Epifluorescence illumination considerations.* A software-controllable light supply or shutters allows for precise control over when the cells are illuminated and with which color light. Software-controlled light emitting diode (LED) offer many advantages as light intensity remains relatively constant over time and wavelengths can be changed very fast when the device is coupled with a dual or multiband pass dichroic mirror and filter set. For example, rapid sequential imaging of mCherry and GFP can be accomplished using a Lumencor Spectra X (Beaverton, OR) with 575/25 nm and 470/22 nm excitation filters coupled with an FF505/626-Di01 dual pass dichroic mirror and an FF01-524/628-25 dual-pass emission filter by Semrock (Rochester, NY) [19]. Alternatively, a mercury arc or xenon illuminator can be used with excitation filters and a mechanical shutter. A shuttered system works well for single color fluorescence imaging, although multicolor imaging requiring mechanical filter wheels or cube carriages are often too slow to provide sufficient temporal resolution. For these types of systems, an optical beam splitter in the emission pathway, such as the Photometrics DV2 (Tuscon, AZ), is needed for simultaneous imaging of two channels [17]. If two color imaging is required, special care must be taken to align the two channels. For a more through description of alignment, *see* Picco et al. [17].

5. If strains contain a plasmid, the strains must be maintained under selection.

6. For most strains, cultures will be grown to near stationary phase overnight, but strains with low viability may still be growing in log phase. If a strain with low viability is to be imaged, inoculate the starter culture to a higher $O.D._{600}$ so there will be sufficient cell growth by the time the other strains are ready for imaging.

7. The ConA will adsorb to any glass it contacts so spreading the ConA around the coverslip can result in a larger area of cell adherence. However, leave a ~3 mm border around the outside of the coverslip where the slide grease will be used to stick the coverslip to the slide.

8. If a deviation from this protocol requires imaging cells in alternative media, cells should still be adsorbed in imaging media. We find that YPD and potentially other rich media inhibit cell adsorption.

9. After washing, the cells should be attached and separated from one another by several µm. If loose cells are observed or cells are on top of one another, cells should be washed more vigorously. If there are few to no cells visible, wash the cells less vigorously. Many endocytic mutants cause cells to clump. For cells with severe endocytosis defects, it may be necessary to search the slide for isolated cells as adjacent cells can contribute to background fluorescence.

10. Slides should be imaged within 40 min as the cells may begin to divide and dislodge from the coverslip and cells may begin to deplete the available nutrients.

11. Optimization of the imaging parameters will likely be required for every experiment. Slower acquisition provides greater exposure time at the expense of time resolution and faster acquisition sacrifices brightness. If sequential imaging is used to capture each channel, a slight temporal offset will occur between the two channels. However, data acquired with sequentially imaged channels does not appear to significantly change any conclusions about the spatiotemporal recruitment of endocytic proteins if the total time to image both channels does not exceed 1 s [19].

12. To minimize photobleaching, we recommend avoiding epi-fluorescence illumination before starting the image capture. There is much less photobleaching when dim transillumination is used to find and focus on the cells to be imaged.

13. Digital images should be encoded with sufficient bit depth to capture meaningful differences in signal intensity. A TIFF image is a generic uncompressed file format that is readable by almost all image analysis software. If compression is required for data storage, be sure to use a lossless compression algorithm.

14. When images are recolorized by ImageJ/FIJI, selecting magenta and green rather than red and green will make images more accessible for individuals with some forms of color blindness [20].

15. Endocytosis is polarized and endocytic patches are most concentrated in the bud of each cell. This can make it difficult to differentiate between individual patches. Therefore, we recommend quantifying patches in the mother cell so neighboring patches do not interfere with the analysis.

16. Most endocytic mutations also eliminate patch polarization to the daughter cell.

Acknowledgments

NSF-MRI grant DBI-1126711 and Lawrence University Biology Department. NTU SUG M4081533 and MOE2016-T2-1-005(S) to Y. Miao, Singapore.

References

1. Brodsky FM (2012) Diversity of clathrin function: new tricks for an old protein. Annu Rev Cell Dev Biol 28:309–336. https://doi.org/10.1146/annurev-cellbio-101011-155716

2. Boettner DR, Chi RJ, Lemmon SK (2012) Lessons from yeast for clathrin-mediated endocytosis. Nat Cell Biol 14:2–10. https://doi.org/10.1038/ncb2403

3. Weinberg J, Drubin DG (2012) Clathrin-mediated endocytosis in budding yeast. Trends Cell Biol 22:1–13. https://doi.org/10.1016/j.tcb.2011.09.001

4. Kaksonen M, Toret CP, Drubin DG (2005) A modular design for the clathrin- and actin-mediated endocytosis machinery. Cell 123:305–320. https://doi.org/10.1016/j.cell.2005.09.024

5. Lewellyn EB, RT a P, Hong J et al (2015) An engineered minimal WASP-myosin fusion protein reveals essential functions for endocytosis. Dev Cell 35:281–294. https://doi.org/10.1016/j.devcel.2015.10.007

6. Skruzny M, Brach T, Ciuffa R et al (2012) Molecular basis for coupling the plasma membrane to the actin cytoskeleton during clathrin-mediated endocytosis. Proc Natl Acad Sci U S A 109:E2533–E2542. https://doi.org/10.1073/pnas.1207011109

7. Sun Y, Martin AC, Drubin DG (2006) Endocytic internalization in budding yeast requires coordinated actin nucleation and myosin motor activity. Dev Cell 11:33–46. https://doi.org/10.1016/j.devcel.2006.05.008

8. Boulant S, Kural C, Zeeh J-C et al (2011) Actin dynamics counteract membrane tension during clathrin-mediated endocytosis. Nat Cell Biol 13:1124–1131. https://doi.org/10.1038/ncb2307.Actin

9. Janke C, Magiera MM, Rathfelder N et al (2004) A versatile toolbox for PCR-based tagging of yeast genes: new fluorescent proteins, more markers and promoter substitution cassettes. Yeast 21:947–962. https://doi.org/10.1002/yea.1142

10. Lee ME, DeLoache WC, Cervantes B, Dueber JE (2015) A highly characterized yeast toolkit for modular, multipart assembly. ACS Synth Biol 4:975–986. https://doi.org/10.1021/sb500366v

11. Longtine MS, Iii AMK, Demarini DJ, Shah NG (1998) Additional modules for versatile and economical PCR-based gene deletion and modification in saccharomyces cerevisiae. Yeast 961:953–961

12. Kaksonen M, Sun Y, Drubin DG (2003) A pathway for Association of Receptors, adaptors, and actin during endocytic internalization. Cell 115:475–487. https://doi.org/10.1016/S0092-8674(03)00883-3

13. Schneider CA, Rasband WS, Eliceiri KW (2012) NIH image to ImageJ: 25 years of image analysis. Nat Meth 9:671–675

14. Michelot A, Grassart A, Okreglak V et al (2013) Actin filament elongation in Arp2/3-derived networks is controlled by three distinct mechanisms. Dev Cell 24:182–195. https://doi.org/10.1016/j.devcel.2012.12.008

15. Pawley JB (2006) Fundamental limits in confocal microscopy. In: Pawley JB (ed) Handb. Biol. Confocal Microsc, 3rd edn. Springer, New York, pp 20–42

16. Sun Y, Leong NT, Wong T, Drubin DG (2015) A Pan1/End3/Sla1 complex links Arp2/3-mediated actin assembly to sites of clathrin-mediated endocytosis. Mol Biol Cell 26:3841. https://doi.org/10.1091/mbc.E15-04-0252

17. Kukulski W, Schorb M, Kaksonen M, Briggs J a G (2012) Plasma membrane reshaping during endocytosis is revealed by time-resolved electron tomography. Cell 150:508–520. https://doi.org/10.1016/j.cell.2012.05.046

18. Rines DR, Thomann D, Dorn JF et al (2011) Live cell imaging of yeast. Cold Spring Harb Protoc 2011:1026–1041. https://doi.org/10.1101/pdb.top065482

19. Picco A, Ries J, Nedelec F, Kaksonen M (2015) Visualizing the functional architecture of the endocytic machinery. Elife:04535

20. Landini G, Perryer DG (2011) More on color blindness. Nat Meth 8:891

Chapter 18

Using FM Dyes to Monitor Clathrin-Mediated Endocytosis in Primary Neuronal Culture

Michael A. Cousin, Sarah L. Gordon, and Karen J. Smillie

Abstract

This protocol utilizes lipophilic FM dyes to monitor membrane recycling in real time. FM dyes are virtually nonfluorescent in solution but when membrane bound are intensely fluorescent, combined with the flexibility of different emission wavelengths make these dyes an excellent choice for investigating clathrin-mediated endocytosis, among other membrane trafficking and recycling pathways.

Key words FM dyes, Clathrin-mediated endocytosis, Primary neuronal culture, Live cell imaging, Epifluorescence

1 Introduction

Fluorescent monitoring of presynaptic function is essential since the small size of a typical central nerve terminal precludes real time monitoring by conventional electrophysiological approaches. The monitoring of FM dye loading and unloading is a technique which can directly report membrane recycling in real time at all sizes of nerve terminal. FM dyes are ideal tools to study endocytosis because they are impermeant to membranes and increase their quantum yield 350 fold in a lipophilic environment [1]. The FM dye is applied during and just after stimulation and therefore, following endocytosis, only internalized membrane will be labeled, enabling staining of membrane specifically in an activity-dependent manner [2, 3]. A further advantage of FM dyes is that they report membrane turnover without the need to perturb the system being studied, for example with transfection of exogenous reporters.

The properties of specific FM dyes are conferred by different parts of the molecule (Fig. 1a). The dyes are amphipathic with a hydrophilic, positively charged head group and a hydrophobic tail

---

The original version of this chapter was revised. A correction to this chapter can be found at
https://doi.org/10.1007/978-1-4939-8719-1_19

Laura E. Swan (ed.), *Clathrin-Mediated Endocytosis: Methods and Protocols*, Methods in Molecular Biology, vol. 1847,
https://doi.org/10.1007/978-1-4939-8719-1_18, © Springer Science+Business Media, LLC, part of Springer Nature 2018

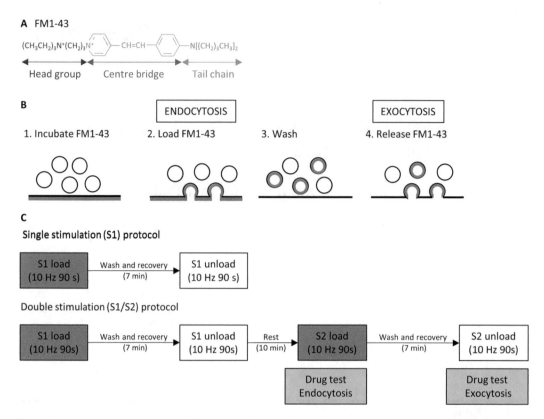

Fig. 1 Overview of the structure of FM dyes and the loading/unloading assay to monitor clathrin-mediated endocytosis. (**a**) Schematic showing the molecular structure of FM1-43 highlighting the head group, center bridging region and the tail chain. (**b**) Schematic of the FM dye assay to monitor synaptic vesicle turnover. Step 1: incubate the FM1-43 with the neurons, labeling the outer leaflet of the plasma membrane. Step 2: stimulate the neurons to evoke synaptic vesicle (SV) exocytosis and subsequent SV endocytosis. During SV endocytosis the FM1-43 will be internalized and the neurons are said to be "loaded." Step 3: wash the neurons to remove any extracellular FM1-43 which was not internalized. Step 4: stimulate the neurons once more to evoke SV exocytosis and "unloading" of the FM1-43. This unloading step acts as a measure of the dye which was initially endocytosed. (**c**) Schematics of the stages of the experiment showing the loading and unloading conditions for a single stimulation (S1) protocol and a double stimulation (S1/S2) protocol. FM dye should also be included in the wash solution for 90 s poststimulation during the loading section of the protocol. The drug incubation points in an S1/S2 protocol are also indicated

linked together with a double bonded bridging region. The head group determines the membrane permeability and has a dicationic charge preventing the molecule passing through the membrane [2]. This ensures that the dye remains in the outer leaflet of the plasma membrane. The length of the hydrocarbon tail determines the affinity of the FM dyes for the membrane bilayer. Dyes with longer hydrocarbon tails, such as FM1-43 have slower departitioning rates than those with shorter hydrocarbon chains, such as FM2-10 and thus FM2-10 is more easily washed off than FM1-43 [2, 3]. The central double bond bridging region is responsible for the spectral properties of the dye: the greater the number of double bonds, the longer the wavelengths required to excite the molecule. Therefore, the double bonded FM1-43 emits green light, whereas the triple bonded FM4-64 emits red light [2, 3].

As stated above, different FM dyes are available with discrete molecular properties [2], allowing experimental design to be tailored to the question being addressed. It is also possible to use fixable versions of the FM dyes, with modified aliphatic amines which can be cross-linked with formaldehyde or glutaraldehyde fixatives [4]. These modified dyes are generally used for end point assays; however, they can also be used for real-time monitoring of membrane recycling if required [4, 5]. The protocol described below (based on Gordon et al. [6]; Ryan and Smith [7]) uses the nonfixable version of FM1-43 but this protocol could be used with any of the other FM dyes, for example FM2-10 or FM4-64.

The main principle behind this assay is that the FM dye is allowed to partition into the external leaflet of the plasma membrane under basal conditions. When the neurons are stimulated, evoking synaptic vesicle exocytosis and subsequent endocytosis, the labeled membrane is internalized (called "loading"). The remaining dye on the outer leaflet of the plasma membrane is then washed away, leaving only the internalized FM dye in the endocytosed compartments. Following a rest period to allow the neurons to recycle and generate fusion competent synaptic vesicles, the neurons can then be stimulated once more to allow release of the internalized dye by exocytosis. This step is designed to unload all of the internalized dye (called "unloading") by applying a maximal stimulus to the neurons known to mobilize the entire recycling pool of vesicles. Therefore, this unloading step can be used to quantify the amount of dye internalized during the first loading stimulation (Fig. 1b). Due to the nature of the assay—using an exocytosis step to initiate endocytosis of the FM dye, it is important to monitor any alterations to exocytosis, since this will impact on the amount of internalized dye and may present as an endocytosis defect. The kinetics of dye release can be used to calculate the speed of exocytosis and therefore be used to delineate between a genuine endocytosis defect and a confounding exocytosis defect. Thus, information on multiple stages of synaptic vesicle recycling can be obtained from one experiment.

The first part of this protocol describes a single stimulation experiment (S1 experiment, Fig. 1c), which can be used for a number of experimental paradigms. For example, this assay can be used to compare synaptic vesicle recycling between different genotypes of culture. It is also useful for examining the impact of overexpression of exogenous proteins of interest or silencing of endogenous proteins on synaptic vesicle recycling following transfection of a DNA construct. Due to the low transfection efficiencies typically seen in primary neuronal culture, this allows for analysis of the transfected neuron but also nontransfected neurons in the same field of view, acting as an excellent internal control.

The second part of this protocol details a double stimulation experiment (S1/S2 experiment, Fig. 1c). This style of experiment can be used to directly compare an experimental variable, for example the effect of adding a drug to the cultures. The first stimulation (S1) acts as the internal control in this instance. The drug can then be added to the culture during the rest period between stimulations and during the second load (Fig. 1c) to examine the effects on endocytosis. Alternatively the drug can be added after the second load and during the second unload to examine effects on exocytosis (Fig. 1c). The second stimulation (S2) therefore acts as the test condition. In this way, the effects of the drugs can be analyzed on the same nerve terminals, thus controlling for variability between cultures and coverslips.

The final part of the protocol details suggested steps for the post hoc analysis of the data collected during the experiments. It explains the importance of normalizing the data to allow comparison between experiments as well as tips on screening the data set with inclusion criteria to ensure meaningful conclusions can be reached. It also explains how to use the data to analyze exocytosis as well as endocytosis.

2 Materials

2.1 Equipment

1. Sealed imaging chamber with embedded platinum wires such as a Warner Imaging chamber (Warner RC-21BRFS).

2. Electrical field stimulator (*see* **Note 1**).

3. Perfusion system, for example a gravity flow system with peristaltic pump (*see* **Note 2**).

4. Inverted epifluorescence microscope with standard fluorescein and rhodamine filters and ×20 or ×40 objective (*see* **Note 3**). A digital camera connected to the microscope plus the associated operating software will also be required.

5. Primary neuronal cells cultured on glass coverslips compatible with the imaging chamber (*see* **Note 4**).

6. Image J (http://imagej.nih.gov/ij/), Microsoft Excel and GraphPad Prism or other similar software to analyze the gathered time series data.

2.2 Solutions and Materials

1. Imaging buffer for cortical/hippocampal neurons: 136 mM NaCl, 2.5 mM KCl, 2 mM $CaCl_2$, 1.3 mM $MgCl_2$, 10 mM glucose, and 10 mM HEPES [pH 7.4] supplemented with 10 μM 6-cyano-7-nitroquinoxaline-2,3-dione and 50 μM DL-2-amino-5-phosphonopentanoic acid (*see* **Note 5**).

2. Imaging buffer for cerebellar granule neurons: 170 mM NaCl, 3.5 mM KCl, 0.4 mM KH_2PO_4, 20 mM TES (*N*-tris(hydroxy-methyl)-methyl-2-aminoethane-sulfonic acid),

5 mM NaHCO$_3$, 5 mM glucose, 1.2 mM Na$_2$SO$_4$, 1.2 mM MgCl$_2$, 1.3 mM CaCl$_2$, pH 7.4.

(If using high potassium solution (*see* **Note 1**) instead of electrical stimulation, high potassium imaging buffer is as above but with the addition of 50 mM KCl and removal of 50 mM NaCl.)

3. FM1-43 dye (Life Technologies): 1 mM in water (100× stock) stored at −20 °C until use (*see* **Note 6**).

4. Vacuum grease.

5. ADVASEP-7 (optional, *see* **Note 7**): 1 mM final concentration [8].

3   Methods

3.1 Single Stimulation Experiment

All steps are carried out at room temperature, unless otherwise stated.

1. Take a coverslip with primary neurons and replace the culture medium with the appropriate imaging buffer. Allow the cells to equilibrate for a minimum of 10 min (*see* **Note 8**).

2. Assemble the imaging chamber (*see* **Note 9**), mount on the microscope and locate a field for imaging using the bright-field (or fluorescence for transfected neurite if required (*see* **Note 10**)).

3. Dilute the FM1-43 in imaging buffer to 10 μM ready for use (*see* **Note 11**) and inject into the imaging chamber to bathe the neurons in the FM1-43 solution (*see* **Note 12**).

4. Stimulate the neurons with 900 action potentials delivered at 10 Hz (*see* **Note 13**) to evoke exocytosis and compensatory endocytosis (and therefore uptake of FM1-43) and continue to incubate the cells in the FM dye solution for a further 90 s (*see* **Note 14**).

5. Wash neurons with imaging buffer for 7 min to remove the noninternalized FM1-43 and to allow the neurons to recover from stimulation.

6. Near the end of the recovery period, check and refocus the microscope if required.

7. Record approximately 20 s baseline before once more stimulating the neurons with 900 action potentials delivered at 10 Hz to release the internalized FM1-43, recording throughout and for 1 min following the end of the stimulation, gathering images every 4 s (*see* **Note 15**). Example images of loaded fields of view prior to the unloading stimulation and following stimulation can be seen in Fig. 2a, b respectively.

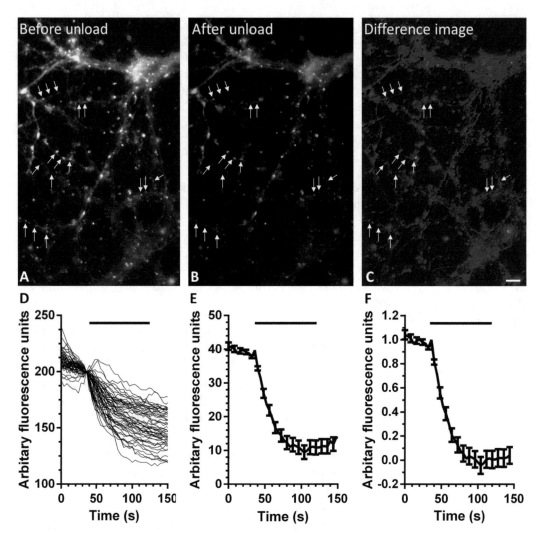

Fig. 2 FM dye assay example data. (**a–c**) Example images showing cortical neurons loaded with FM1-43 using a 10 Hz 90 s stimulation before (**a**) and after (**b**) unloading with a subsequent 10 Hz 90 s stimulation. (**c**) Difference image created by subtraction of the image after stimulation from the image before stimulation. The warmer the color, the greater the amount of dye unloaded. Arrows highlight nerve terminals which undergo unloading of FM1-43. Scale bar represents 10 μm. (**d**) Raw unloading data normalized to the point of stimulation. (**e**) Average unloading data following decay correction ± SEM (n = 1 coverslip with n = 54 nerve terminals). (**f**) Unloading data normalized between 0 and 1 to enable calculation of the time constant of dye unloading ± SEM (n = 1 coverslip with n = 54 nerve terminals). Black bar represents the period of stimulation

Note: If electrical stimulation is not possible, this protocol can be adapted to load and unload the neurons with a high potassium solution (*see* **Note 16**).

3.2 Double Stimulation Experiment

1. Carry out the protocol from **steps 1** to **7** for a single stimulation experiment as above for the first stimulation and pause the image acquisition. This stimulation can act as an internal control for drug studies.

2. Allow the neurons to recover for a minimum of 10 min. If investigating the effects of any drugs on endocytosis, preincubate the neurons with the drug at this point.

3. Dilute the FM1-43 in imaging buffer as before (plus drug if required) and inject into the chamber. Load the neurons with FM1-43 as before, including the poststimulation FM1-43 incubation and then wash away any noninternalized FM1-43.

4. Allow the neurons to recover once more for 7 min. If investigating the effects of any drugs on exocytosis, incubate the neurons with the drug during this time period.

5. Check the focus, restart image acquisition and unload the internalized FM1-43 as before. This second stimulation acts as the test stimulation for drug studies.

3.3   Data Analysis

1. Check the raw time series from the experiment for any focus drift issues and reregister the time series to correct any *xy*-drift (*see* **Note 17**).

2. Place regions of interest (ROI) over areas on the neurite corresponding to nerve terminals and measure the fluorescence at these points over time using the Time Series Analyzer plugin for ImageJ (http://rsb.info.nih.gov/ij/plugins/time-series.html, *see* **Notes 18** and **19**).

3. Normalize the fluorescence signal for each ROI to an arbitrary value at the point of stimulation. This removes the variation due to the starting fluorescence value and allows comparison between ROIs (*see* **Note 20**). An example of the normalized raw data from an experiment is shown in Fig. 2d.

4. Fit an appropriate single exponential to the average trace, using GraphPad Prism for example, to correct for the fluorescence decay due to bleaching if required. This can be done using the baseline frames at the start of the data acquisition and applied to the individual ROI traces (*see* **Note 21**).

5. The individual ROIs can now be averaged to give the total fluorescence decrease. This represents the total amount of dye initially internalized and hence the extent of endocytosis (Fig. 2e). If the data was from a double stimulation experiment, the two stimuli responses can be compared by ratio for example to determine the effect of any drugs tested.

6. Kinetic information on the release of the FM1-43 can also be informative as an indirect measure of exocytosis rate. To do this normalize the start and end points of the unloading stimulus between 0 and 1 and calculate the time constant (tau, Fig. 2f, *see* **Note 22**).

4 Notes

1. Electrical stimulation is the most physiological method to stimulate cells; however, it is possible to use a modified version of this protocol to use high KCl solution (*see* **Note 16**) to evoke a clamped depolarization and initiate exocytosis for loading and unloading the cells [9].

2. When setting up the tubing to connect the perfusion rig to the imaging chamber, an injection line for the FM dye is recommended to reduce the volume of dye solution required. This piece of tubing as well as the tubing connecting the tubing manifold to the imaging chamber should be kept short to also minimize the volume of dye required to reach the chamber. This is important since the FM dyes can quickly degrade in solution at room temperature. A clip to seal off the injection line when not in use is also required.

3. An inverted microscope is required so that the oil objective can be in contact with the bottom of the sealed imaging chamber.

4. For example, if using a Warner Imaging chamber, coverslips smaller than 25 mm in diameter will not completely seal the imaging chamber leading to buffer leaks and larger than 25 mm in diameter coverslips will not fit into the cassette the imaging chamber sits in.

5. A concentrated (10 times) stock of the imaging buffer can be prepared and stored at −20 °C. On the day, an aliquot can be thawed, diluted and warmed to room temperature for imaging.

6. FM1-43 is very sensitive to light and temperature. It is recommended to purchase a pack of small aliquots (100 µg) rather than one larger vial. The lyophilized powder stocks should be stored at −80 °C in a light-proof container until required. Minimize exposure to light when reconstituting. Once reconstituted, the FM1-43 should be aliquoted into single use aliquots and stored at −20 °C in a light-proof container until required. Once an aliquot is thawed for use, it should be kept on ice in the dark until diluted into the imaging buffer for loading of the cells.

7. Depending on the primary cultures used, for example cortical neurons, the background staining can be quite high and so the removal of any noninternalized dye can be improved with the addition of ADVASEP-7 to the wash solution. This binds to the FM1-43 and helps to remove it from the external leaflet of the plasma membrane [10].

8. The equilibration step at the start of the experiment is essential if using cerebellar granule neurons since these neurons are grown in depolarizing media and require to be repolarized before the experiment.

9. To ensure a good electrical contact, at the start of each imaging day, the platinum wires should be gently rubbed with some fine sandpaper to remove any salt deposit. If vacuum grease is required to assemble the imaging chamber, for example a Warner imaging chamber, the wires should also be cleaned carefully to ensure that they are free from grease. During assembly vacuum grease should be applied with just sufficient to seal the coverslips. An excess of vacuum grease can contribute to z-drift during imaging due to the coverslip slowly settling over time. Once assembled and connected to the perfusion rig it is good practice to check for any buffer leaks before mounting onto the microscope.

10. Choosing the field to image is critical for the success of an experiment. The field should not be too densely populated with neurites, since this makes post hoc analysis difficult. Areas where the neurites have formed bundles should also be avoided due to limitations of identifying individual neurites which is especially important when looking at transfected neurites.

11. If the tubing lines are kept relatively short, for example approximately 10 cm of tubing with an internal diameter of 1.6 mm, 1.5 ml of diluted FM1-43 solution should be sufficient to prime the line and incubate the cells in dye.

12. Injection of the FM1-43 solution into the chamber must be done carefully to avoid injecting any air bubbles into the chamber. Air bubbles reduce the volume of buffer in the chamber and if sufficiently large can prevent even conduction of the field potential across the cells.

13. The stimulation conditions selected to load and unload the FM dye are important depending on the question being addressed because use of high intensity stimulation (greater than 20 Hz) conditions will evoke other membrane recycling pathways, such as activity-dependent bulk endocytosis, as well as clathrin-mediated endocytosis [11]. If higher intensity stimulation protocols are required and the effect only on clathrin-mediated endocytosis is to be investigated, use of FM2-10 at the correct concentration will only label membrane being retrieved via CME [12].

14. Incubation with FM dye following the end of stimulation ensures that all membrane endocytosed is labeled since membrane retrieval via clathrin-mediated endocytosis persists following stimulation [11].

15. Occasionally, spherical fluorescent bubble-like structures are seen in the field of view following incubation of the neurons with FM1-43 solution. These are seen when the FM1-43 has degraded, which can occur when FM1-43 remains in the tubing lines for too long. To avoid this, re-prime the dye

injection line following a break from imaging. If this continues to be a problem, the stock may have expired and fresh FM1-43 should be made up.

16. Adapted protocol to use a high potassium solution (50 mM) instead of electrical stimulation to load and unload the dye. In this case, the FM1-43 should be diluted in high potassium imaging solution and should be perfused onto the neurons for 30 s to load the cells. FM1-43 should also be included in the wash solution for 90 s to label endocytosis occurring post-stimulation. To unload the neurons, the high potassium imaging solution should be perfused onto the cells in two sequential pulses each for 30 s, approximately 100 s apart. Please note that a clamped depolarization will also evoke other membrane recycling pathways in addition to clathrin-mediated endocytosis—*see* **Note 13**.

17. Reregistering the raw time series data before analysis is important to ensure that the ROIs remain on the selected nerve terminal throughout the movie. Drift in x and y can be corrected using the Stack-reg plugin for ImageJ (http://fiji.sc/StackReg). Z-Drift cannot be corrected and so any time series with this issue cannot be accurately analyzed and should be discarded.

18. Identifying functional nerve terminals is a skilled procedure since there can be a lot of punctate staining within the field of view and there can be a strong decay in the fluorescent signal due to photobleaching. One useful method is to make a subtraction image using the images directly before and after stimulation. This will create an image showing regions where dye was unloaded in an activity-dependent manner, facilitating visualization of the responsive ROIs over the general bleaching of the fluorescence signal (Fig. 2c).

19. If transfected neurons are being used, ROIs can be placed on both transfected neurites and also untransfected neurites in the same field of view (It is recommended to do this in a separate file to avoid confusion). This is a useful internal control for such experiments and can be used for comparison between coverslips if the response of the transfected neurite is expressed as a proportion of the untransfected neurites.

20. Following normalization it is important to examine the data and remove any ROIs which are not responding to stimulation from further analysis. A response to stimulation is defined as a stimulation-dependent decrease in fluorescent signal. This will underestimate an inhibition but it is important to have confidence that the data included in the analysis is sound and from responding, healthy neurites. However, a complete block of clathrin-mediated endocytosis will fail to load a significant quantity of FM dye and therefore will not display a stimulation-dependent decrease.

21. If the data is from a double stimulation experiment, it is likely that the decay from the first stimulation will be different to the second stimulation. In this instance, independent exponential decays should be used for the two parts to the experiment.

22. Normalizing the start point of the average fluorescence drop to 1 and the end point to 0 means that any differences in total amount of dye internalized (and subsequently released) will have no impact on the kinetic rate of release calculated, assuming that labeled and unlabeled vesicles intermix randomly. This enables exocytosis to be compared between neurons displaying control amounts of endocytosis and those displaying inhibited endocytosis for example.

Acknowledgment

This work was supported by the Medical Research Council (grant number G1002117).

References

1. Henkel AW, Lubke J, Betz WJ (1996) FM1-43 dye Ultrastructural localization in and release from frog motor nerve terminals. Proc Natl Acad Sci U S A 93:1918–1923

2. Betz WJ, Mao F, Smith CB (1996) Imaging exocytosis and endocytosis. Curr Opin Neurobiol 6(3):365–371

3. Ryan TA (2001) Presynaptic imaging techniques. Curr Opin Neurobiol 11(5):544–549

4. Renger JJ, Egles C, Liu G (2001) A developmental switch in neurotransmitter flux enhances synaptic efficacy by affecting AMPA receptor activation. Neuron 29:469–484

5. Rea R, Li J, Dharia A et al (2004) Streamlined synaptic vesicle cycle in cone photoreceptor terminals. Neuron 41:755–766

6. Gordon SJ, Leube RE, Cousin MA (2011) Synaptophysin is required for Synaptobrevin retrieval during synaptic vesicle endocytosis. J Neurosci 31(39):14032–14036

7. Ryan TA, Smith SJ (1995) Vesicle pool mobilization during action potential firing at hippocampal synapses. Neuron 14(5):983–989

8. Kim SH, Ryan TA (2009) Synaptic vesicle recycling at CNS synapses without AP-2. J Neurosci 29(12):3865–3874

9. Ryan TA, Reuter H, Wendland B et al (1993) The kinetics of synaptic vesicle recycling measured at single presynaptic Boutons. Neuron 11(4):713–724

10. Kay AR, Alfonso A, Alford S et al (1999) Imaging synaptic activity in intact brain and slices with FM1-43 in *C. elegans*, lamprey and rat. Neuron 24(4):809–817

11. Clayton EL, Evans GJ, Cousin MA (2008) Bulk synaptic vesicle endocytosis is rapidly triggered during strong stimulation. J Neurosci 28:6627–6632

12. Clayton EL, Cousin MA (2008) Differential labelling of bulk endocytosis in nerve terminals by FM dyes. Neurochem Int 53:51–55

Correction to: Clathrin-Mediated Endocytosis: Methods and Protocols

Laura E. Swan

Correction to:
Laura E. Swan (ed.), *Clathrin-Mediated Endocytosis:*
Methods and Protocols, Methods in Molecular Biology, vol. 1847,
https://doi.org/10.1007/978-1-4939-8719-1

This book was inadvertently published with the incorrect title as Clathrin-Mediated Endoytosis: Methods and Protocols. This has now been corrected throughout the book to Clathrin-Mediated Endocytosis: Methods and Protocols.

The updated online version of this book can be found at
https://doi.org/10.1007/978-1-4939-8719-1

Laura E. Swan (ed.), *Clathrin-Mediated Endocytosis: Methods and Protocols*, Methods in Molecular Biology, vol. 1847,
https://doi.org/10.1007/978-1-4939-8719-1_19, © Springer Science+Business Media, LLC, part of Springer Nature 2018

INDEX

Laura E. Swan (ed.), *Clathrin-Mediated Endocytosis: Methods and Protocols*, Methods in Molecular Biology, vol. 1847,
https://doi.org/10.1007/978-1-4939-8719-1, © Springer Science+Business Media, LLC, part of Springer Nature 2018

Printed in the United States
By Bookmasters